# BRAINSTORMS

*and*

# MINDFARTS

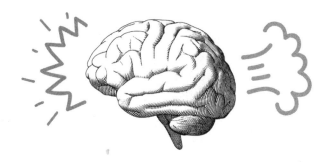

# BRAINSTORMS

*and*

# MINDFARTS

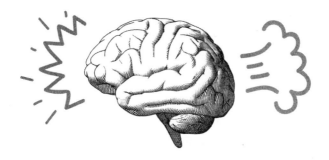

## The Best and Brightest, Dumbest and Dimmest Inventions in American History

TOM CONNOR *and* JIM DOWNEY

RUNNING PRESS
PHILADELPHIA

Running Press
Hachette Book Group
1290 Avenue of the Americas, New York, NY 10104
www.runningpress.com
@Running_Press

Printed in the United States of America

First Edition: May 2021

Published by Running Press, an imprint of Perseus Books, LLC, a subsidiary of Hachette Book Group, Inc. The Running Press name and logo is a trademark of the Hachette Book Group.

The Hachette Speakers Bureau provides a wide range of authors for speaking events. To find out more, go to www.hachettespeakersbureau.com or call (866) 376-6591.

The publisher is not responsible for websites (or their content) that are not owned by the publisher.

The photographs on pp. 5, 25, 31, 37, 51, 81, 111, 113, 153, 175, 203 copyright Getty Images

Print book cover and interior design by Rachel Peckman.

Library of Congress Control Number: 2020947057

ISBNs: 978-0-7624-7244-4 (hardcover), 978-0-7624-7246-8 (ebook)

LSC-C

Printing 1, 2021

JUL 1 6 2021

*To every kid with a little daydream and
every grown-up with a big idea, we dedicate this
collection to prove that anything is possible.*

# CONTENTS

# BRAINSTORMS

*and*

# MINDFARTS

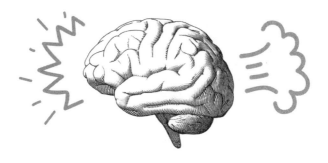

# Introduction: How Great Inventors and Innovators See the World

Passion and insatiable curiosity mark the inventive mind.

Throughout American history, the great inventors and innovators have shared these essential traits. Gazing far into the future, they saw the products and services that one day would transform not only the country but the world. The Founding Fathers were innovators and entrepreneurs. While passionate about creating this new thing called a democracy, our Founding Fathers, such as Benjamin Franklin, were also driven to change the way humans lived and worked—to complete everyday tasks faster, more easily, and more efficiently. As with all visionaries, they simply liked seeing their brainy ideas become reality, no matter how myopic and ridiculous—how asinine and flatulent, if you will—a good number of them might be!

Let us not neglect, then, the otherwise forgotten mad or sad geniuses and dreamers whose obsessions drove them to file patent applications for products destined for un-greatness. We can only guess at the Rube Goldberg–like workings of their minds, but, as a class, they seem to share a number of characteristics: (1) they thought they were geniuses; (2) they were convinced that the patent applications they were filing would not only be granted but would make them a fortune; and (3) in spite of all previous evidence to the contrary, they believed their spouses would finally stop saying, "You're such a fool!"

Take, for example, the Dog Toy (US636069391), which is a fake stick (see page 170). Far be it from us to call anyone else a fool—after all, we thought it was a good idea to write about patents like the Dog Toy!—but the "invention" seems less the product of pure genius than one of pure laziness.

And yet the funny thing about patents and products like this is that sometimes they pop up at precisely the right moment for the world to wake up one morning and say, "That's *ridiculous!* I *want* one!"

Take, for a more inspiring example, the Slinky (US2415012A). In 1943, mechanical engineer Richard James accidentally knocked a coil of wire springs off a shelf and onto the top of a nearby flight of stairs, then watched as the springs "walked" gracefully down the steps. When the Slinky, as he called it, appeared in the window of Gimbels department store in Philadelphia over Christmas in 1945, four hundred of the toy novelties sold out within minutes. To date, over 350 million Slinkys have sold (see page 174).

But back for a moment to the pioneering inventors who gave America its reputation for innovation and many of the world's most remarkable inventions.

While working in his brother's Philadelphia print shop, a young man by the name of Benjamin Franklin, arguably the progenitor of American inventions and patents, was fascinated by the effect of machines on design and production, and their potential both to improve on existing technologies and to create what hadn't been seen before, like bifocal lenses, or refrigeration, or scuba fins.

"In many cases, great inventors and innovators are trying to solve a problem," says Fairfield, Connecticut-based patent attorney Paul Fattibene, who has filed more than a thousand patents for clients in his career. "It's a matter of finding a solution to that problem that others haven't come up with yet . . . It's a matter of taking a fresh approach to an existing problem and, hopefully, a simpler, less expensive solution."

"Not everybody is willing to sit down and think through the solution to a problem," notes Sean Patrick Suiter, founder of the intellectual property firm Suiter Swantz in Omaha, Nebraska. "I don't really know a lot of successful inventors who have 'Eureka!' moments, although the popular press portrays it that way. Yet there's certainly that moment when there's this conception of a fully formed idea of how to solve a problem, or the identification of the source of a problem, but these are individuals who have been thinking about the problem for a while."

A constant flow of ideas seems fundamental to success when it comes to innovation. Thomas Edison reportedly filled 3,500 notebooks with jottings and drawings for new gadgets by the time of his death in 1931 at age eighty-four.

As of 2018, the US Patent & Trademark Office had granted its ten millionth patent. But with over 500,000 applications being filed annually, fewer than half of these applicants will be granted patents and far fewer still—an estimated 1 percent of patent applications—will realize commercial success, according to the Patent & Trademark Office. Some are flawed by mistakes or missing details; others are too ridiculous to take seriously; still others are simply ahead of their time.

This, however, will probably have little dampening effect on the current crop of inventors, innovators, and entrepreneurs. Inventors don't think in terms of success and failure. When an idea doesn't pan out, rather than abandoning it, they're more likely to file it in the back laboratories of their minds and move on to the next idea. "Thomas Fetidson" (it is said that Edison went days without bathing) reportedly tried over 900 variations of the light bulb before finding the one that worked. His explanation? That's how many steps the pioneering technology required.

Inventors don't necessarily "invent" so much as they look at existing problems with fresh eyes. As historians have pointed out, inventors also don't see the world as it is—they see the world as it can be and are continually challenging accepted norms.

What's more, pioneering inventors rarely create something from scratch. Rather, they often stand upon the shoulders of earlier thinkers and inventors. Just as Leonardo da Vinci is said to have based his drawings for a flying machine on an ancient Chinese toy, the Wright brothers drew inspiration from French engineer Paul Cornu who, thirty-seven years earlier, built a model helicopter powered by nothing more than rubber bands (Cornu may well have been inspired by da Vinci). So when Igor Sikorsky was credited with inventing the first functional helicopter in 1940, it was less a matter of a fresh invention than one of improving on earlier models, often found in nature.

George de Mestral, for example, came up with the idea of Velcro (see page 52) after disengaging, with some difficulty, the tiny hooks of cockleburs off his pant legs and dog fur from hunting in the Swiss mountains.

Excluded to a great degree from the patent process, and thus missing from these pages, are inventions by women and African-Americans, mirroring the history and culture of the country. Yet, for centuries, free as well as enslaved blacks contributed countless innovations that made life easier for their masters and employers, who usually took credit for their inventions (see also Lonnie G. Johnson and the Super Soaker, page 178). Meanwhile, women like Vesta Stoudt, who conceived of duct tape (page 36) and actress Hedy Lamarr, the brains behind a secret communications system that aided American allies at the start of World War II (page 54), succeeded where men hadn't.

For some inventors, the seeds of a new technology and product come from the simple if hubristic desire to make life personally better. Sony cofounder Masaru Ibuka wanted to be able to listen to operas during his frequent trans-Pacific plane trips, and gave the task to one of his engineers. The result was the Sony Walkman, the predecessor to the iPod. Following its debut in 1979, it was enormously successful.

Mark Zuckerberg is said to have been looking for a way to help fellow Harvard undergrads connect with one another and ended up inventing Facebook (though the Winklevoss twins have taken issue with this, saying that they were the true brains behind the idea).

But while problem-solving on an epic scale is often the result of selfless acts of wanting to help humanity—Alexander Graham Bell first became interested in the science of sound because both his mother and his wife were deaf—ego drives many inventors and innovators: They want to see their ideas become reality simply because those ideas are *their* ideas.

Failure doesn't factor into the equation; if anything, it drives them harder to experiment, to think outside the box. Believing in themselves and their vision, they simply aren't willing to give up on what they see as something the world one day will—or should—thank them for.

At the same time, there's a companion trait that seems unique to pioneering thinkers and innovators: the unremitting desire to prove disbelievers and naysayers wrong.

Apple was well on its way to developing a prototype for the iPad in 2007 when Steve Jobs, founder and at the time CEO of the company, abruptly ordered his team to stop work on the project. According to Scott Forstall, a member of the Apple team assigned to the project, Jobs was driven to create the iPhone first after being taunted by a Microsoft executive who said his company was working on a similar project and was going to rule the world.

Jobs had this advice for novice innovators: "Don't let the noise of others' opinions drown out your own inner voice. And, most important, have the courage to follow your heart and intuition." To intellectual property attorney Sean Suiter's point about egocentric personalities, it isn't likely that "others' opinions" would have drowned out Steve Jobs, but we know what he meant.

By exploring some of the greatest pioneering patents, and the inspiration and perspiration behind them, contemporary innovators and entrepreneurs have open access to the minds and imaginations of their creators, and gain valuable insight into the alchemy by which ideas become new technologies, products, and services.

They may also learn something from the not-so-great—the non-pioneering, the no-way-could-their-patents-ever-be-considered-even-remotely-pioneering—inventors; namely, that an idea can't be worth much until it is tested, the inventor applies for a patent for it and that patent is granted, and the world judges its value.

For all the patents that appear in print here, the courage of the creators' convictions is worth at least their weight in ink.

# PATENTS PAST, PRESENT, AND PENDING

# CHAPTER ONE

—

# Business and Industry

Innovation and entrepreneurism define US business and industry. Behind every iconic product and brand borne of the American imagination are the patents that drive and validate our insatiable quest for newer, faster, better. The inventions in this section tend to be grounded in the here and now work world, but that doesn't mean they can't inspire us to dream of the future. The fun here is for us to use *our* imaginations to picture the office worker of 1867 marveling at the ingenuity of the paper clip, and that same worker entering the modern office to find something called a 3-D printer spitting out plastic versions of the same. Let's follow him through these pages and to lunch at the diner, where Flippy the Burger-Flipping Robot will slap a burger on his bun.

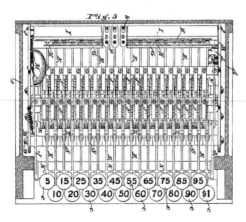

Drawing submitted as part of the application
for the first patented cash register.

## Dynamite (improved explosive compound)

US78317
Inventor: *Alfred Nobel*
Patent Granted: *May 26, 1867*

Behind all great foundations are family fortunes, and behind most of those are world-changing innovations or pioneering patents and products.

Alfred Nobel is best-known today for endowing the annual Nobel Prizes for advances in physics, chemistry, physiology and medicine, literature, economics, and peace. Each year, the Nobel Foundation presents winners with a gold medal, a diploma, and over $900,000 in prize money to help them build their careers. But to Nobel, a scientist and entrepreneur, destructive had appeared more lucrative than constructive: A substantial portion of his fortune was made by discovering a better, safer way to blow stuff up.

Nobel and his younger brother, Emil, worked with their father manufacturing nitroglycerine, a highly unstable liquid explosive that can spontaneously detonate. After Emil and other workers died in an explosion in the family's factories, Alfred was determined to make the explosive safer to handle by mixing nitroglycerine with silica to make a paste that could be formed into rods and inserted in holes drilled into material to be detonated.

Dynamite (from the Greek *dynamis*, meaning "power") changed work across multiple industries—mining, construction, highway, and railroad building—and boxy blasting machines with push-down plungers became a common prop in early cartoons, Westerns,

# A Half-Serious Note to Readers

The US Patent & Trademark Office has a standardized way of presenting the patents it grants, which are published on a weekly basis. Each patent is numbered, beginning with US 1 in 1790 (see page 30). Since then, the USPTO has granted over six million patents, with more than 300,000 new applications filed every year.

The titles of the patents that appear in this book often differ from the names under which they're filed. That's because when patents become products, they usually acquire names created by marketers or consumers. What we refer to today as the iPhone, for example, was filed under "Electronic Device."

In addition to numbers, patents are assigned codes like A, for a utility patent issued before January 2, 2001, or A2, issued after that date. ("Utility" patents cover the creation of new or improved and useful products, processes, or devices.) You might also see a "D" following the "US" that precedes a patent number and identifies a patent covering the ornamental design of a useful item.

Here, in brief, are two other terms you'll see. An "abstract" is simply a short summary of a patent application. "Abandoned" refers to a patent application that is no longer pending because the application didn't comply with one or more USPTO requirements.

As for the process by which we've included or excluded the inventions found in the following pages, we don't know what to tell you: We've only included those we've found interesting or amusing or both, plus the pioneering patents and products on which contemporary culture relies. These are, for the most part, simply observations and opinions filtered through the often distorted lenses of our sensibilities and for which we can make no apology nor take any responsibility: We're not patent attorneys, or intellectual property professionals, or experts on anything. We're not even really adults (as the title of this book should have made clear).

In other words, now that you've hopefully purchased this book, you're basically on your own. So good luck!

and war movies. Just as important, it launched from the '70s sitcom *Good Times*, the groundbreaking Jimmie Walker catchphrase, "Dyn-o-mite!"

### 3-D Printer (apparatus for production of three-dimensional objects by stereolithography)

US4575330A
Inventor: *Charles W. Hull*
Patent Granted: *March 11, 1986*

You would think we'd be satisfied by printing pages at home without having to build a press, form words from metal letters in a tray, and apply ink to those words by hand. But apparently not.

In 1986, humans realized that what we've *really* needed all along was not just being able to print letters and term papers, but to print handguns and pizzas! That realization coincided, believe it or not, with an invention created the usual way: out of frustration with the status quo.

Three years earlier, Chuck Hull, an engineer and inventor, was working one night in the lab behind his house in Colorado. The company he worked for had been using UV light to harden layers of plastic veneer on furniture. But when a prototype for any new product was needed, it would take months for its parts to be designed, produced, and delivered.

Hull's bright idea? If enough layers of photopolymers could be "printed" on top of one another, then etched and shaped with ultraviolet light, they would harden to form three-dimensional objects. The process has since been compared to building a sandwich by stacking slices of food and condiments between slices of bread. Hull called it "stereolithography" and the result was not a fast-order sandwich service but the first iteration of 3-D printers, which use computer-aided design, or CAD, to digitally design a desired object.

Although Hull had envisioned the future, he couldn't have foreseen what would come flying out of his invention. Beyond guns and pizzas, to date 3-D printers have "printed" a plethora of useful and wacky objects (not unlike the range of patents in this book!), including buildings, bikinis, a wall-climbing robot, a drivable sports car, skin, a model of a fetus before birth, and, wackiest of all, a 3-D printer!

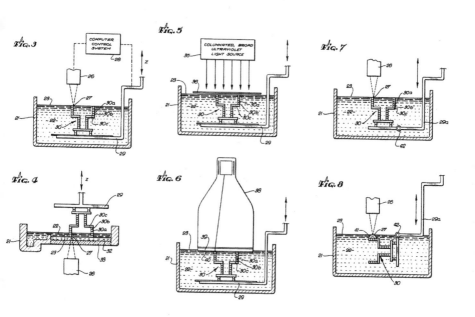

## Cash Register

US271363
Inventors: *James Ritty, John Birch*
Patent Granted: *January 30, 1883*

It's one thing to make money; it's another to keep track of it. And when moola mysteriously goes missing, some are inspired not only to find out where it went but to devise a way so that it doesn't happen again.

In the case of this famously patented machine, desperation more than necessity was the mother of invention.

That's the situation James Ritty found himself in back in 1871. A saloon owner in Dayton, Ohio, Ritty's popular joint was regularly packed but still bled dough, which was kept in a cigar box behind the bar. Suspecting his bartenders of thievery, he changed them frequently, to no avail. According to numerous accounts, in order to escape his troubles, Ritty sailed to Europe and onboard observed a piece of machinery that counted the revolutions of the ship's propellers.

Back in Ohio, he applied the concept to his predicament and, with pal John Birch, eventually came up with what he called Ritty's Incorruptible Cashier—a clunky beast with rows of keys denoting amounts from five cents to a dollar and a crankshaft that turned a counter to track the day's sales.

Heavy and ornate, brass cash registers from Ritty's day through the early twentieth century can be found in saloons, antiques stores, and private collections today. Fortunately, contemporary registers are electronic or computerized, saving cashiers from having to count change in their heads or on their fingers.

Currently, nearly half a million brick-and-mortar retail stores are operating across the United States, with half again as many cash registers. They have changed size, shape, construction, and functionality since Ritty's day, but there's one thing that hasn't changed for store owners as well as for the nation's manufacturers: They bring in cash. *Ka-ching!*

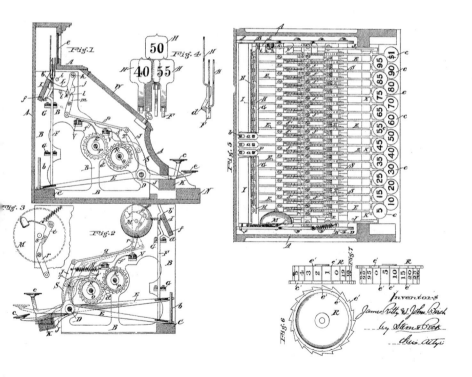

## Pencil with Eraser

US19783
Inventor: *Hymen Lipman*
Patent Granted: *March 30, 1858*

Digital natives might think life's mistakes will simply vanish with the click of the "delete" key, or the "clear history" function on their devices. Digital immigrants like us know better.

Before the major shift in communications that this patent represents, if we were hand-writing something important in pencil—a shopping list, a note to a girl across the aisle in school—and made a mistake, we had to put the pencil down, go look for an eraser, rub out the mistake, leaving rubbery rubbings all over the paper and desk, then pick the pencil up and try again. So much wasted motion and time!

There had to be a better way—and there was! It just took an entrepreneur and owner of a stationery store in Philadelphia named Hymen Lipman to seize the day and usher in a huge technological leap forward from the plain pencil. His ingenious idea? Stick an eraser on the other end of the pencil from the point. (The rubber tip was also a great improvement on the sixteenth-century version: balled-up bread!)

Actually, Lipman wasn't the first to conflate pencil-drawn marks and their obliteration, according to an account posted on Haaretz, the Israeli news service. But he was the first to patent and profit from them.

In 1862, he smartly sold his patent for a whopping $100,000 (the equivalent of $2 million today) to Joseph Reckendorfer, a local businessman. A decade later, after A.W. Faber (now Faber-Castell, one of the oldest and largest pencil and pen companies in the world) brought out a similar product, Reckendorfer sued on grounds of patent infringement.

In a rare turnaround, the suit reached the US Supreme Court, which in finding that both the pencil and the eraser were existing technologies and therefore the combination was nothing new, invalidated Hymen Lipman's patent.

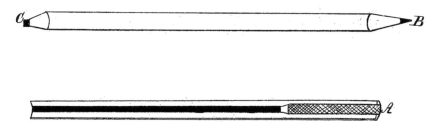

*H. L. Lipman.*
*Pencil & Eraser.*
*Nº 19,783. Patented Mar. 30, 1858.*

## Self-Sharpening Pencil

US549952
Inventor: *Frederick E. Blaisdell*
Patent Granted: *November 19, 1895*

Yeah, we're old school—go ahead and smirk! But for centuries the pencil was *the* high-tech tool in human communication. And when your tablets and mobile phones go down, and when your pens run out of ink, you may hold the humble yellow sticks of wood and graphite in gratitude and esteem.

According to "The History of the Pencil" on Pencils.com (behind every object lies not just history but a website!), the Romans invented the pencil's precursor—the stylus. Some were made of metal that left marks on papyrus, a precursor to paper; others were made of lead and, later, a mineral called graphite. In England in the sixteenth century, scribblers used graphite sticks wrapped in string before someone had the bright idea of sticking the stuff in hollowed-out sticks of wood. The first wood pencils in the New World were made in Massachusetts in the early 1800s, where Henry David Thoreau is believed to have made his own to write *Walden*. (Vladimir Nabokov is said to have written *Lolita* and all his novels in pencil, as did John Steinbeck, who reportedly used up as many as sixty a day!)

For more on this most utilitarian of devices—and there's a lot more!—you might want to read Henry Petroski's *The Pencil: A History of Design and Circumstances*, published in 1992.

As with so many inventions, many hands work toward product development, and this is true of the pencil as well. Nicolas-Jacques

Conte, a scientist in Napoleon Bonaparte's army, received French Patent 32, building on the work of earlier pencil makers. But here, main patent bragging rights go to Philadelphian Frederick Blaisdell, whose self-sharpening pencil unwraps when worn to expose a fresh tip.

Let's see you do that with a malfunctioning keyboard!

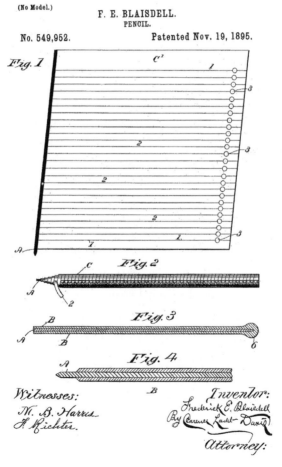

(No Model.)

F. E. BLAISDELL.
PENCIL.

No. 549,952.

Patented Nov. 19, 1895.

## ATM (credit card currency dispenser)

US3761682
Inventors: *Thomas R. Barnes, George R. Chastain, Don C. Wetzel*
Patent Granted: *September 25, 1973*

Withdrawing money from a bank account without stepping inside a bank once seemed inconceivable. In fact, it took a long time for the concept to catch on. As is often the case with patents, however, inconvenience ultimately leads to invention, and more than one inventor is often involved.

In 1968, Don Wetzel was standing in line in a Dallas bank when it occurred to him that maybe he didn't have to. As vice president of product planning at Texas-based Docutel, Wetzel had been involved in the company's development of automated baggage-handling equipment. If a machine could handle bags, why couldn't it handle money?

Others, it turned out, had already answered that question. In 1939, Luther Simjian field-tested his version of an automated teller machine for what is now Citicorp, which shelved the venture after six months, due to low consumer demand. Inventors in Japan and Scotland developed similar machines. But it was Wetzel, with fellow Docutel employees Tom Barnes and George Chastain, who applied for and were granted a patent for what former US Federal Reserve chairman Paul Volcker reportedly called the "only useful innovation in banking."

Today, as foot traffic inside brick-and-mortar banks has declined, automated teller machines are doing a lot more than spitting out cash. Synced with other ATMs around the world, they perform a host of banking functions, from depositing cash and checks to checking balances and transferring money between accounts.

An estimated 400,000 ATMs currently dot the American landscape and some three million litter the planet. Instead of waiting impatiently in line inside a bank, customers now wait impatiently in line at an ATM for the clicking to stop and the cash to slide out. *Hurry up!*

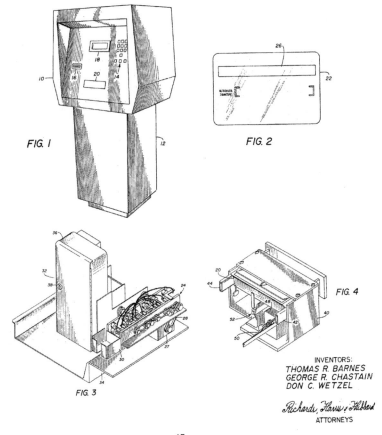

FIG. 1

FIG. 2

FIG. 3

FIG. 4

INVENTORS:
*THOMAS R. BARNES*
*GEORGE R. CHASTAIN*
*DON C. WETZEL*

ATTORNEYS

## Flippy, the Burger-Flipping Robot
## (robotic kitchen assistant for preparing food items in a commercial kitchen and related methods)

US20200121125
Inventors: *David Zito, Sean Olson, Ryan W. Sinnet, Robert Anderson, Benjamin Pelletier, Grant Stafford, Zachary Zweig Vinegar, William Werst*
Applicant: *Miso Robotics, Inc.*
Patent Granted: *April 23, 2020*

Timing counts in the patent-granting process, as in all human endeavors, and the approval of the patent for this invention comes during a pandemic, when cleanliness, physical distancing, and cost efficiency can mean life or death for countless eateries.

Robots have become familiar fixtures in industrial workplaces. But with this automated kitchen assistant, they're beginning to make an appearance in diners and fast-food joints across the country.

Developed by Miso Robotics, Inc., an artificial intelligence (AI) solutions company, Flippy consists of a flexible robotic arm and controller, a sensor assembly with multiple cameras, and a processor that uses AI to perform fry-cook duties based on orders, recipes, kitchen equipment, and camera data. Its processor is synced with a restaurant's point-of-sale system so that food starts being cooked as soon as an order is input at the register. They're collaborative robots, the company says, designed to work with humans, not replace them. (It's interesting to note that, though efficiency is part of the plan, it took eight men to invent one robot!)

Beyond a one-time, upfront paycheck, the advantages of robotic line cooks versus humans are speed of service, consistency of orders, and reduced food waste. And given the repetitive, stressful nature of the job, unless Flippy's had a tough day and overindulges in lubricant out back, he's likely to be a highly dependable employee.

The Miso robot was scheduled to start work at a White Castle restaurant in Chicago in September 2020. Should a customer have a complaint about an order, a simple H.I. (human intelligence) solution exists: "Talk to Flippy!"

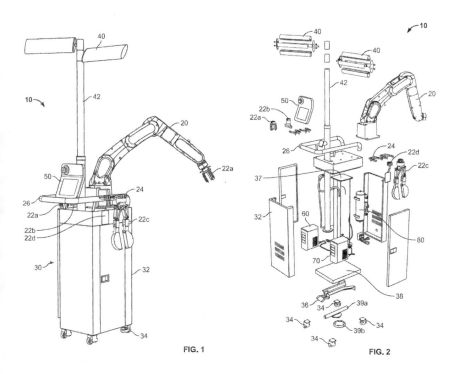

FIG. 1                    FIG. 2

## Bitcoin Mining Machine (optimized SHA-256 datapath for energy-efficient high-performance Bitcoin mining)

US10142098
Inventors: *Vikram B. Suresh, Sudhir K. Satpathy, Sanu K. Mathew*
Current Assignee: *Intel Corporation*
Patent Granted: *November 27, 2018*

Our first thought when coming across this term on the digital frontier was "Really?" And our second thought: "No, *really?*" I don't know what we were expecting—some steel beast with a giant drill for a snout, or miners loaded into a cart descending deep into the mine shaft, but no. Those exist in the actual world and this in the virtual world.

It helps, of course, to know what Bitcoins are: They're digital currencies traded on a peer-to-peer network outside the oversight of banks or exchanges. Transactions within the Bitcoin currency system are recorded and kept in a public online ledger or Blockchain. (There's even an ICO—an initial coin offering—which is the cryptocurrency industry's equivalent of an IPO, an initial public offering, as a way to raise funds.)

Now that that's clear, Bitcoin mining refers simply (thanks to Cointelegraph.com) to a process of record-keeping, through computer processing power, to confirm every Bitcoin transaction on the Blockchain. And it's used to distinguish legitimate Bitcoin transactions from attempts by banditos (sorry) to re-spend the currency that has already been spent elsewhere.

Yet you might not know this from the patent application's abstract, which explains that this Bitcoin mining system "includes a processor to construct an input message comprising a plurality of padding bits and a hardware accelerator, communicatively coupled to the processor, comprising a first plurality of circuits to perform a stage-1 secure hash algorithm (SHA) hash based on . . ."

You get the idea. For our money, we'll stick with cash.

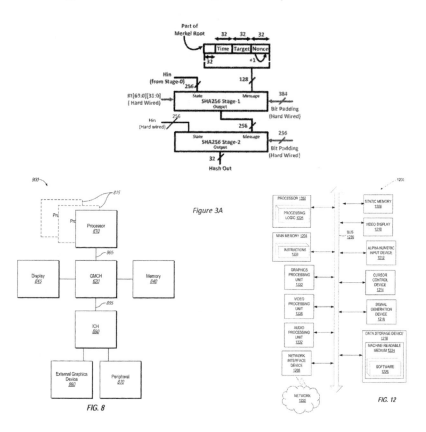

Figure 3A

FIG. 8

FIG. 12

## Typewriter (improvement in typewriting machines)

US79265

Inventors: *C. Latham Sholes, Carlos Glidden, Samuel W. Soule*

Patent Granted: *June 23, 1874*

Mention this once-disruptive technology to youngsters today and you're likely to hear, "What's a typewriter?"

Yet for more than a century before the advent of the digital keyboard, these clunky machines changed the way students wrote papers and the majority of adults communicated in print.

Johannes Gutenberg started the ball rolling in 1440, as you may recall, with the invention of a printing press with movable metal type. For the first time in history, it permitted the printing of thousands of sheets of vellum a day, replacing the need for monks and scribes to produce a few dozen pages with rudimentary pens and ink. During the 1450s, it's estimated that the press produced 180 copies of perhaps the most famous books ever printed—1,286 pages, bound in two volumes, of the Gutenberg Bible.

Over the next four centuries, hundreds of movable type machines came and went in an attempt to personalize the technology. Then, in 1874, three Milwaukee, Wisconsin, natives received a patent for an "Improvement in Type Writing Machines."

Compared with today's silent, effortless keyboard, the early models were cumbersome: Lettered keys, when pressed, moved long metal arms with raised letters at the ends. Those, in turn, struck an inked ribbon that ran from a spool on the left, across the front of the keywell and onto a second spool. A sheet of paper was inserted into a roller, and when a line of typed letters reached the far edge of the page, a typist hit a metal bar to return the carriage, or the top of the machine, back to the opposite side. It's tiring just writing about the process!

Obsolete though they may be, remnants of them can be found on digital devices today. Hundreds of typewriter fonts, like Courier and Olivetti, offer a retro look. Apps like Qwertick and Noisy Typer re-create the clickety-clack of manual keys. But the most influential remnant of this patented device is the QWERTY keyboard layout, named for the first six keys of the top row, which remains the standard layout on any number of electronic device keyboards. Another remnant that's all but obsolete: People like us who type with one finger!

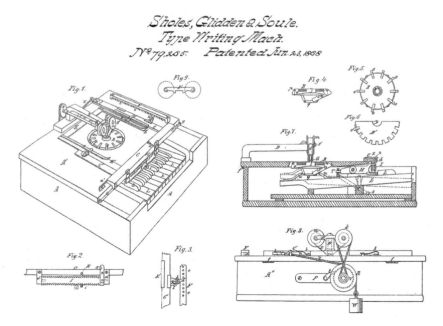

## Paper Clip

US64088

Inventor: *Samuel B. Fay*

Patent Granted: *April 23, 1867*

Imagine a business district in any big city at the turn of the twentieth century. The miracle of electricity has recently illuminated workplaces and homes, and is poised to power a bright future. Alexander Graham Bell's telephone has now changed daily life forever by allowing husbands, wives, teenagers, and in-laws to hang up on each other at will. But for now, the streets are still alive with the clatter and commotion of horse-drawn vehicles.

Let us now enter a typical office workplace of the time. We'll find rows of clerks at heavy wooden desks, topped with green ink blotters, rubber stamps of every variety, and, occasionally, one of those newfangled typewriters. Compare this charmingly antiquated scene to the sleek, spare, ergonomically efficient cube farm of today, and you'll find these two disparate environments have nearly nothing in common, except for the one indispensable item that has literally transcended time: the paper clip.

First patented by Samuel Fay in 1867, the original design changed over the years, finally evolving, through various iterations, into the clip shape we know today as the Gem, a masterpiece of design simplicity, introduced without a patent in 1892. (By the late 1800s, a flood of paper clips patents were filed, including W.W. Cole's, shown in the illustration at right.)

Today, twelve billion paper clips are sold in America every year, and, by dint of their sheer ubiquity, have been ingeniously pressed into service as salt shaker un-cloggers, reset button pushers, and eyeglass hinges. You've got plenty of them in that drawer in the kitchen alongside the dead AA batteries.

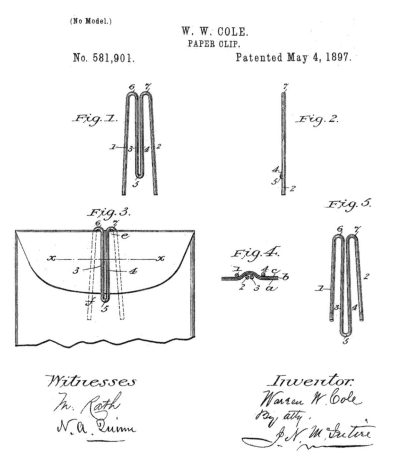

(No Model.)

W. W. COLE.
PAPER CLIP.

No. 581,901.                    Patented May 4, 1897.

Fig. 1.    Fig. 2.

Fig. 3.    Fig. 4.    Fig. 5.

Witnesses
M. Rath
N. A. Quinn

Inventor.
Warren W. Cole
By atty.
B. N. McIntire

## Combine Harvester

US9793X
Inventor: *Hiram Moore*
Patent Granted: *June 28, 1836*

For hundreds of years, America has proudly held the title of Bread Basket to the World, and for good reason. The Great Plains, the huge arable swath across the middle of the country, has reliably produced enough wheat, oats, rye, barley, corn, and soybeans to feed generations of Americans, and a goodly portion of the rest of the world's population as well.

In 1800, nearly 90 percent of the entire US population was employed working the land. Today, just 2 percent of people are farmers. One of the main factors driving this massive societal change was the development of automated machines, which made every worker vastly more efficient and productive.

If there's one machine that can be credited with much of this technical progress, it's the combine harvester. As its name implies, the combine is designed to effectively harvest a variety of grain crops by "combining" three separate operations—reaping, threshing, and winnowing—into a single process.

Almost every kid's American history book credits Cyrus McCormick with the invention of the reaper, counting him alongside Eli Whitney and Thomas Edison as the men who drove America's industrial progress. But that's not the way the residents of Climax, Michigan, a small town in Kalamazoo County, see things. They think the story has a different protagonist.

A certain Hiram Moore, one of the village's founders in eastern Kalamazoo County, had an idea for a horse-drawn machine that would combine the reaper and the thresher operations. The problem was that neither of these machines had been perfected. This left the door open for Moore's competitor, Cyrus McCormick, whose ethics at the time were pretty shaky when it came to industrial espionage.

It seemed that every time Moore completed a set of blueprints for his New York foundry, McCormick had already come up with a copy and so beat Moore to the patent office. He used Moore's own concepts, eventually garnering all the credit, wealth, and fame. Hiram Moore didn't take this treatment lying down. He sued, and the bitter legal fight cost McCormick some $40,000.

But, in the end, McCormick got the wheat and Moore got the chaff.

A Cyrus McCormick reaper circa 1859.

## Self-Destructing Document and E-Mail Messaging System

US7191219B2

Inventors: *Howard R. Udell, Cary S. Kappel, William Ries, Stuart D. Baker, Greg M. Sherman*

Current assignee: *Hanger Solutions, LLC*

Patent Granted: *March 13, 2007*

Here's a useful tool for the Age of Email! How often have we wished we'd never sent that self-incriminating message or file? ("Too often" would be our answer.) Actually, today dozens of systems, devices, and apps serve the human desire to alter the past, but back in 2007 just one patent stood out.

The system as described in the patent application's abstract automatically destroys sent documents or emailed messages at a predetermined time by means of a virus attached or embedded in them.

Original paperwork relating to the system is said to have been filed by executives or representatives of Purdue Pharma, the privately held pharmaceutical company headquartered in Connecticut, as early as 1997. The 2007 patent, filed in 2002, replaced the 1997 patent.

The dates are of interest because the year before, Purdue began marketing OxyContin, the highly addictive opioid that has since been cited in some two thousand lawsuits for deceptive marketing connected to deaths from the painkiller.

According to a 2019 ABC News report, the earlier patent's existence came to light during a deposition given by Richard Sackler, Purdue Pharma's former chairman and president. So did the fact that, in May 2007, less than two months after the current patent was granted, Purdue pled guilty to misleading patients, doctors, and regulators about the possibility of addiction and death due to OxyContin's abuse.

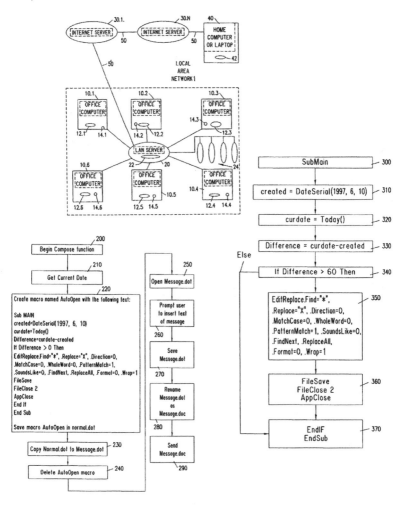

## CHAPTER TWO

# Science and Technology

More than perhaps any other section in this book, the inventors represented here stand on the shoulders of the pioneering giants and inventors of the past. Without Galileo Galilei and Nicolaus Copernicus, there would be no modern-day telescope; without Leonardo da Vinci, perhaps no Orville and Wilbur. What's also interesting here are the peeks into history that many patents provide. The Process for Making Potash, an early fertilizer and the very first patent, was granted in 1790 and signed by Thomas Jefferson. In stark contrast are the scary glimpses into the future from recently granted patents for creatures like Robotic Bees, which may soon buzz in a summer sky near you. But neither should diminish the culture-changing innovation that is Duct Tape, used these days to fix everything under the sun except perhaps ducts. The real genius when it comes to the product? American teenagers. Who else would think of constructing prom dresses and tuxes out of the stuff?

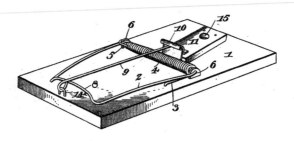

"Build a better mousetrap, and the world will beat a path to your door,"
Ralph Waldo Emerson advised. In 1894, William C. Hooker did.

## Process for Making Potash

US1
Inventor: *Samuel Hopkins*
Patent Granted: *July 31, 1790*

We don't hear much about potash these days. Basically, it's fertilizer. But back in 1781, when the country was an agricultural nation and people actually grew what they ate, the stuff was a hot commodity and in serious demand. (It would take nearly another century before the Great Atlantic & Pacific Tea Company—the now-defunct A&P—opened the first national grocery store chain.) So big a deal was potash that an entire cottage industry grew up around leaching potassium from wood ash and boiling in it large cast-iron pots— hence the cleverly named by-product!

One reason for referencing the product 230 years later is because Samuel Hopkins, a Philadelphia inventor, developed a new process for making potash and received the first patent ever issued in the United States. Another reason for writing about it now is that Hopkins embodied the American entrepreneurial spirit: Seeing the need for growing more crops faster, he not only improved upon the existing process, but had the foresight to protect his big idea by filing a patent application.

President George Washington and then-Secretary of State Thomas Jefferson, both farmers before moving to the swamplands of Washington, DC, thought the potash-making process worth protecting, too. They signed the application, granting Hopkins the patent and launching an international market for potash. According to John H. Lienhard IV, a professor of mechanical engineering and history at the University of Houston, and creator of the radio program and the online series "The Engines of Our Ingenuity," Hopkins's "process greatly improved the yield of potash as well as its purity. For the next seventy years, America was the world's potash producer."

And thus from those humble potassium ashes rose the greatest agricultural economy on the planet.

## Radio

RE11913

Inventor: *Guglielmo Marconi*

Patent Granted: *June 11, 1901*

The invention of radio communication is generally attributed to Guglielmo Marconi in the late nineteenth century. But it took many decades and the hardest work of the best minds to make it possible for teenagers to blast Death Metal in Mom's car on the way to their soccer game.

The world had accepted wired telegraphy, which, although mystifying to most, proved itself invaluable as its tendrils spanned the globe, drove world finance, and gave instant gravitas to the phrase, "You've got a telegram!"

But, of course, "good enough" wasn't good enough to keep scientists and engineers from positing the idea that the wires could be eliminated, thus creating a wireless telegraph. This would take the discovery of electromagnetic waves, including radio waves, by Heinrich Rudolf Hertz who, after fifty years of research, finally proved a theory of electromagnetism.

The development of radio waves into a communication medium did not immediately follow, as the idea of transmitting radio waves some distance didn't seem to be worth the effort of building a communication system based on this thesis. And the idea of paid advertising wasn't even an idea yet. Enter Guglielmo Marconi, who developed the first apparatus for long-distance radio communication. On December 23, 1900, the Canadian inventor

Reginald Fessenden became the first person to send audio by means of electromagnetic waves, successfully transmitting over a distance of about one mile. Six years later on Christmas Eve 1906 he became the first person to make a public radio broadcast.

Later that evening, the phrase "I don't like the sound of that guy's voice!" made its way into the lexicon.

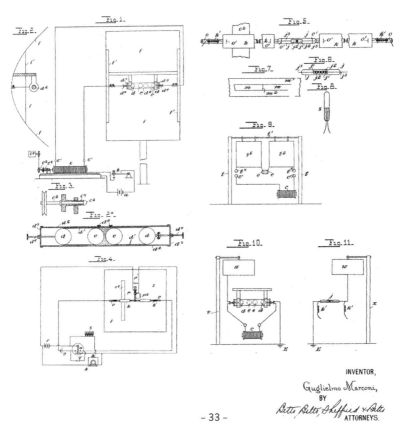

No. II,913.

Reissued June 4, 1901.

G. MARCONI.

TRANSMITTING ELECTRICAL IMPULSES AND SIGNALS AND APPARATUS THEREFOR.

(Application filed Apr. 1, 1901.)

INVENTOR,

Guglielmo Marconi,

BY

*Betts, Betts, Sheffield & Betts*

ATTORNEYS.

## Microscope

US883868A

Inventor: *Joseph H. Ford*

Patent Granted: *April 7, 1906*

High school students in biology classes everywhere can thank Hans Lippershey for their first up-close look at pond scum—and the protozoans swimming around in it—under the lens of his invention. Actually, they can thank him and the scores of lens makers responsible for the modern telescope.

Like the telescope, humans' insatiable curiosity led inventors to look outward and inward in their quest for discovery and knowledge. Some accounts have Galileo Galilei perfecting the first device called a microscope in 1309. But it may well have been Hans Lippershey, who first applied for a patent for his version of the telescope a year earlier (Remember? There may be a quiz at the end of this!), who also invented the first microscope. If true, Lippershey, as the first micro-macro man, might have looked ahead and behind at the same time!

What Joseph Ford did in filing the first US patent was offer improvements to compound microscopes—also known as biological microscopes, used to view specimens not visible to the naked eye—in order to "improve the means for illuminating the field in such instruments and to increase the magnifying power without increasing the tube length of the instrument and without loss of intensity of light."

Between 1853 and 1915, according to Antique-microscopes.com, there were 130 or more patents for the instruments or improvements on the instrument. And according to Justia.patents.com, between

1994 and March of 2020, some two thousand patents were filed for variations on the basic invention.

High school students today can curse all of them for bringing pond scum to squirming, wriggling life. But guess what? The protozoans are looking at the kids and thinking the exact same thing!

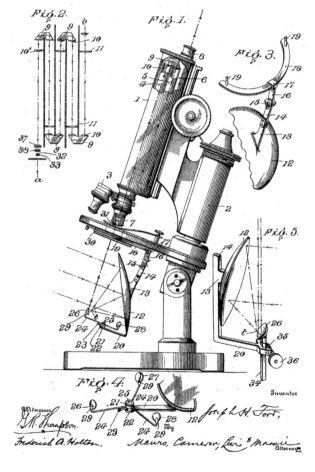

## Duct Tape (waterproof tape)

US5162150A
Inventors: *Frank Nason Manley, Elwood Paul Wenzelberger (on an idea proposed by Vesta Stoudt)*
Assignee: *Johnson & Johnson*
Patent Granted: *February 8, 1944*

A rip in your jeans? No problem. The sole coming away from the rest of your loafer? No worries. Wrap your house securely in advance of a hurricane? A day at the beach. Unroll the tape!

Versatility marks many inventive products, and few inventions have proven as versatile as duct tape. But in 1939, patriotism and a mother's love were what inspired this particular product. It is also worth noting that it took a woman—a rarity in the world of patents, though not in innovation—to see that her brainchild survived.

At the start of World War II, Vesta Stoudt had been packing boxes of ammunition in a factory in Illinois when she realized that soldiers would have a hard time unsealing them in the heat of battle. With two sons in the military, she went home and began experimenting with a waterproof adhesive cloth that was strong but easily torn. When her superiors at work failed to support her idea, Stoudt went over their heads.

"Dear Mr. President," she wrote Franklin Delano Roosevelt, and attached a sketch of her idea. "Please," she wrote in closing, "do something about this at once; not tomorrow or soon, but now."

Within weeks, the military approved Stoudt's invention and commissioned Johnson & Johnson to develop and mass-produce a similar cloth tape with strong adhesive backing (soldiers referred to it as "duck tape," likening it to water rolling off a duck's back).

Now, American teenagers are putting the product to peacetime use: They're making wallets, belts, bracelets, key wristlets, earrings, cell-phone cases, hats, shoes, skirts . . . you name it. There's even a national Stuck at Prom contest, sponsored by Duck Brand duct tape, with awards of $10,000 each for the best prom tux and prom dress made entirely of the brand's tape.

Stoudt's idea stuck.

---

**Camera (method of taking likenesses by means of a concave reflector and plates so prepared as that luminous or other rays will act thereon)**

---

US1582A

Inventor: *Alexander S. Wolcott*

Patent Granted: *May 8, 1840*

---

If it weren't for this nineteenth-century patent, social media would be very different, if it were there at all. Without images of family and friends, landscapes and sunsets, vacations and adventures—without likenesses of ourselves posted again and again—we would engage with one another face-to-face, and when that wasn't possible, write actual letters!

As is often the case, a long line of experimenters made Alexander Wolcott's 1840 patent possible. The desire to capture a moment in time goes back to the earliest cave drawings. From the fifth century BCE to the sixteenth century CE, according to numerous sources, images were projected onto walls or other surfaces through boxes called pinhole cameras or camera obscura.

In Paris in the early nineteenth century, then, Joseph Nicéphore Niépce captured images on bitumen-coated surfaces, a process later refined by Louis-Jacques-Mandé Daguerre, whose "daguerreotypes" were sheets of copper coated with silver iodide and exposed to the light. In Rochester, New York in the late 1880s, George Eastman produced paper film and created the Kodak camera with a fixed focal lens. Compact cameras like the Leica appeared after World War I,

Polaroid cameras in the 1960s, and, in 1975, Kodak's Steve Sasson introduced the first digital cameras.

Alexander Wolcott simply got to the US Patent & Trademark Office first. A dentist and inventor, his patent for a daguerreotype mirror camera enabled him and others to take the life portraits that took as little as five minutes of sitting time. The device used a concave reflecting mirror, similar to those found in celestial telescopes, and Daguerre's chemical process. With John Johnson, Wolcott opened what is believed to be the world's first portrait studio, in New York in 1840, the same year his patent was granted.

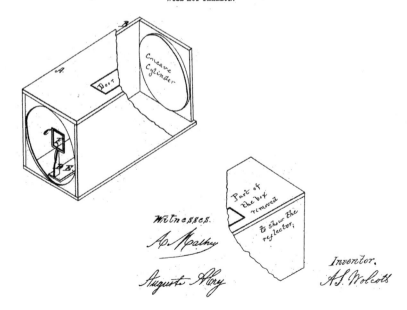

No. 1,582.                                    PATENTED MAY 8, 1840.
A. S. WOLCOTT.
METHOD OF TAKING LIKENESSES BY MEANS OF A CONCAVE REFLECTOR
AND PLATES SO PREPARED AS THAT LUMINOUS OR OTHER RAYS
WILL ACT THEREON.

## Telephone

US174465
Inventor: *Alexander Graham Bell*
Patent Granted: *March 7, 1876*

The ubiquitous telephonic greeting "Hello!" is credited to Thomas Alva Edison, who apparently wasn't satisfied with the 1,093 patents of his own, and so had to horn in on Alexander Graham Bell's glory.

The very first spoken words via telephone, however, were from Bell himself. On March 10, 1876, he picked up the clunky device he'd invented and called his assistant Thomas Watson, who sat waiting for the momentous call in the next room. That Watson may have been able to hear his employer's voice through the wall shouldn't diminish the historic moment. When Bell said, "Mr. Watson—come here—I want to see you," he ushered in one of the most transformative inventions and perhaps the most valuable patent in history.

Bell became interested in the transmission of sound because his mother and his wife were both deaf. Driven by the rapid expansion of the telegraph in the 1870s, inventors like Bell sought to address the problem of scalability—a method with which to send multiple signals across one wire. Bell's ingenious solution was to electronically translate the infinitely variable tones of the human voice into electronic signals that would be reconstituted into recognizable words and phrases on the receiving end.

Submitted on February 14, 1876, the application for a patent was granted a mere three weeks later. In less than a year, the new device made business history, when seven shareholders formed the Bell

Telephone Company. In 1878, Bell wrote, "I believe, in the future, wires will unite the head offices of telephone companies in different cities, and a man in one part of the country may communicate by word of mouth with another in a distant place."

Alexander Graham Bell had the telephone in mind, of course, and in that respect his note was prescient—to an extent. He just didn't know yet about a kid from the future named Steve Jobs.

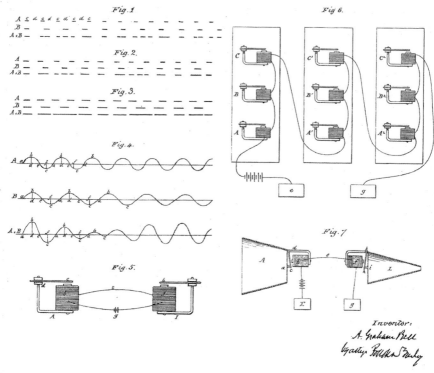

---

**iPhone (electronic device)**

---

USD618677S1

Inventors: *Steve Jobs (plus twelve Apple team members)*

Patent Granted: *June 29, 2010*

---

Billions of people today know what this pioneering device looks like because billions of people own one (including 77 percent of Americans). But in March 2006, when Apple filed the original patent application for the iPhone, all it appeared to be was an unadorned rectangle that might have been drawn by a child: One of the illustrations resembles a juice box with a popsicle stick protruding from the top.

On the application for its patent, the thing wasn't even called an iPhone. Steve Jobs named it an "electronic device," no doubt intentionally to keep it as generic and vague as possible.

Sure, numerous smartphones preceded Apple's innovative idea—IBM's Simon Personal Communicator, launched in 1992, for example—but the elegance of its functionality, combined with Jobs's exquisite design and packaging sense, have made the iPhone so iconic that it has dominated a global industry.

According to Silicon Valley legend, the twelve-man Apple team assigned to develop the iPhone had no idea what the ultimate product would look like or, more importantly, how Jobs planned to make it work. Over the next decade, in fact, hundreds of additional patent applications were filed to expand on the initial concept. Yet the written descriptions accompanying the original application—page after page densely packed with directions and details—provides a deep look into the circuitry of Jobs's and lead Apple designer Jony Ive's brains.

When the pioneering device hit the market on June 29, 2007, it quickly transformed the world as we knew it, allowing users to connect with one another while ignoring face-to-face conversations.

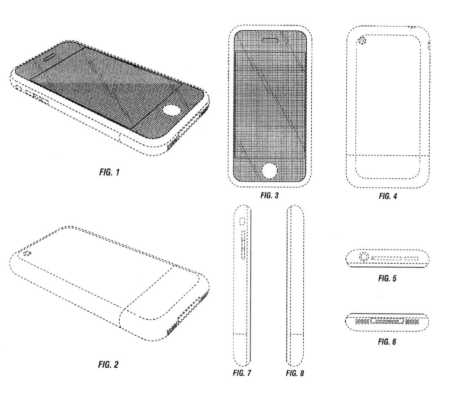

FIG. 1

FIG. 3

FIG. 4

FIG. 2

FIG. 7

FIG. 8

FIG. 5

FIG. 6

## Find My iPhone (apparatus and method for determining a wireless device's location after shutdown)

US201630281113A1

Inventors: *Alexander D. Schlaupitz, Joshua G. Wurzel, Ahmad Al-Dahle*

Assignee: *Apple, Inc.*

Patent Granted: *April 23, 2012*

You pat your pockets. Your dig in your purse. Many times. But it's not there. Your cell phone! You've misplaced or lost it, and you're freaking out!

Apple feels your pain. Or their gain. Either way, marketers at the company figured that since they patented a device that everybody uses, why not patent a second device to find the first device, which everybody loses?

Find My iPhone was the answer. The new technology enables the device to enter a sleep-like state and includes circuitry that powers it up at timed intervals, then sends location data to you via an email message or a text (ideally to another device!). Additionally, the technology allows you to lock the device remotely with a passcode and to display a message with a number where you can be reached. Any data stored on your iPhone will be secured.

But what if your iPhone has been stolen? The inventors thought of this, too. (Hey, they work at Apple! What do you think they do all day? Look for their missing phones like we do?) When powering back up, the device would allow only essential components to be activated, with the sound and display panel kept off so as not to alert the thief that its rightful owner is in hot pursuit. Hopefully, he or she wouldn't receive a text to that effect, either!

Since 2012, the company has filed numerous patent applications relating to Find My iPhone—including twenty-one on the same day! Other companies have also patented devices to find misplaced or stolen devices. At last check, Apple reported that it was looking to combine this device and Find My Friends into a single app that could ultimately track virtually anything, plus identify the location of family members and close friends. And if those people don't want to be found? A device for that service may well be on the way.

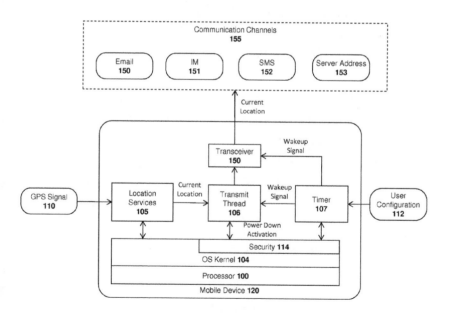

## Robotic Bees (officially filed as systems and methods for pollinating crops via unmanned vehicles)

US15/697106
Applicant: *Walmart*
Patent Filed: *March 8, 2018*
*Patent Pending*

In March 2018, Walmart, the multinational food and retail megacorp, filed for a patent to create robotic bees—technically called "pollination drones"—that would use sensors and cameras to detect and pollinate crops, like real bees, except without the buzz and also without dropping like flies from overwork.

The Walmart application isn't the only one hoping to replace actual bees. As scientists search for solutions to colony collapse disorder and massive die-offs of honeybees, which pollinated an estimated one-third of harvested food, at least five other companies or organizations have filed applications for agricultural drones, including one that could hypothetically attack pests.

In 2013, researchers at Harvard began developing robobees that could dive and swim underwater. At Delft University of Technology in the Netherlands, a team of scientists reproduced the complex wing motion patterns and aerodynamics of the fruit fly in hopes of eventually creating swarms of bee drones to pollinate crops when bee-bees are gone.

But where Harvard-educated robobees can't be remotely controlled, Walmart's can. If the patent is granted and the company brings its bees to market, bumblebots could appear in any number of future scenarios:

surveilling Walmart stores and stinging desultory employees; shooting down competitor bee drones; taking over American agriculture, then other supermarkets, then the world food supply.

As pending patents often reveal, the future can look really scary.

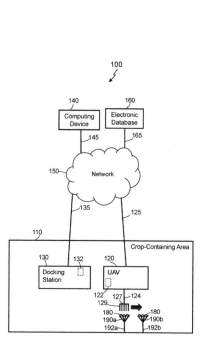

FIG. 1

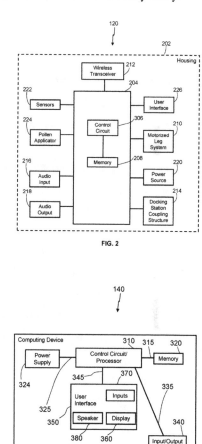

FIG. 2

FIG. 3

## Spring-Loaded Mousetrap

US528671
Inventor: *William C. Hooker*
Patent Granted: *November 6, 1894*

The earliest, and perhaps still the best, mousetrap came with paws with claws, needle-like teeth, and whiskers for navigating in the dark. Cats began earning this distinction some four thousand years ago, but it took until 1894 for William C. Hooker, an Illinois resident and no friend to rodents, to file the first patent for a mechanical mousetrap.

"A simple, inexpensive and efficient trap adapted not to excite the suspicion of an animal"—that is how Hooker described his death machine, which consists of a wooden base, a spring-loaded bar, and a trip to release it. Actually, the spring-loaded "animal trap," as he called it, was preceded by James M. Keep's Royal No. 1, which caught rodents in cast-iron jaws, and was succeeded in 1898 by British inventor James Henry Atkinson's Little Nipper that employed a weight-activated treadle to trip the release.

Whatever the method of execution, it's been curtains for the critters ever since.

"Build a better mousetrap, and the world will beat a path to your door," Ralph Waldo Emerson is alleged to have said. Inventors around the world have certainly tried! Since Hooker's day, more than 4,400 mousetrap patents have been granted in dozens of categories, including Electrocuting and Explosive, Choking or Squeezing, and thirty-six others. According to the Smithsonian Institution's National

Museum of American History, more patents have been awarded to mousetraps than any other invention.

Captive audiences to this entry may want to visit the 150-exhibit mousetrap museum in Bedwas, Gwent, Wales, where the Procter Brothers factory has been manufacturing the Litter Nipper for more than a century.

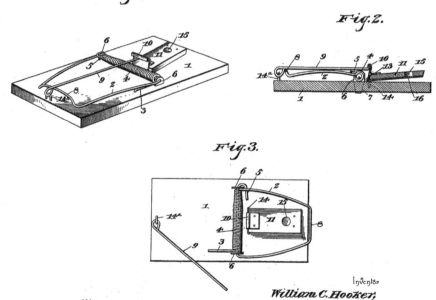

(No Model.)

W. C. HOOKER.
ANIMAL TRAP.

No. 528,671.

Patented Nov. 6, 1894.

Fig.1.

Fig.2.

Fig.3.

Inventor
William C. Hooker;

By his Attorneys:

Witnesses

> ## Insulin (extract obtainable from the mammalian pancreas or from the related glands in fishes, useful in the treatment of diabetes mellitus, and a method of preparing it)
>
> US1469994
> Inventors: *Frederick Banting, Charles Best, James Collip*
> Current Assignee: *University of Toronto*
> Patent Granted: *October 9, 1923*

The discovery of any drug that treats formerly incurable diseases is worthy of inclusion in any book on patents. Insulin is simply one of them. But its development is representative of the painstaking steps and multiplicity of players involved in a pharmaceutical's long journey to the US Patent & Trademark Office, and, ultimately, to the marketplace.

Insulin treats diabetes, a disease that occurs when glucose, or sugar, in the blood is too high. Prior to its discovery, diabetic patients suffered horribly from a host of diseases—heart, kidney, and liver ailments—and were typically given a year, at most, to live following diagnosis. Many lost limbs as well.

In 1889, two German researchers first suggested that substances in the pancreas of dogs might lead to a cure for the disease, according to the American Diabetes Association. Twenty years later, an English physiologist, Sir Edward Albert Sharpey-Schafer, identified the missing chemical in the glands of patients suffering from diabetes. He named it insulin from the Latin *insula*, or *island*, for the isolated chemical found in a healthy pancreas.

Finally, in 1923, three scientists at the University of Toronto extracted insulin and tested it successfully on animals. The first human to receive the new treatment was a fourteen-year-old Canadian boy who was close to death. The first trial failed but the second didn't, and he recovered his health.

That year the scientists were awarded the patent for the drug. To make insulin readily available to anyone who needed it, they sold the patent to the University of Toronto for $1. In 1923 they won the Nobel Prize in Medicine.

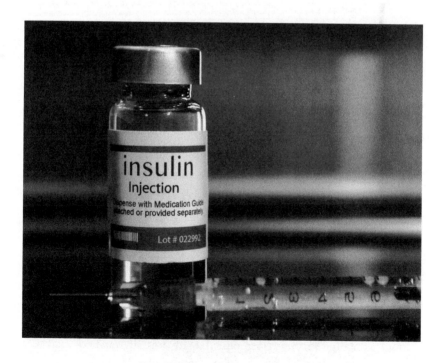

## Velcro

US3009235
Inventor: *George de Mestral*
Patent Granted: *September 13, 1955*

Today, there are still a few people on the planet who haven't heard the distinctive sound of a Velcro fastener being pulled apart, and those few are likely still working on a reliable way to start a fire. To say it's everywhere is an accurate remark. From lunar landers to football helmets, from backpacks to bootstraps, Velcro has become the everyday solution for temporarily but securely attaching items together that otherwise wouldn't have anything to do with each other.

Like so many of history's greatest inventions, the legendary fastener was born as a way to mimic a natural occurrence. It seems that a Swiss engineer named George de Mestral was on a hike with his dog in Switzerland's Jura mountains in the 1940s when he noticed the tiny hooks of a plant called the cocklebur had attached themselves to the weave of his trousers and his dog's coat so stubbornly that it required some serious effort to remove them, much to the chagrin of his pooch.

Later, under a microscope, he studied the hook and loop mechanics of the cocklebur and soon began a quest that would result in the 1955 patent for a Velvet Type Fabric and Method of Producing the Same. The fastener consisted of two components: a fabric strip with tiny hooks that would mate with another fabric strip with smaller loops, attaching securely until pulled apart. Initially made of canvas, the fastener was eventually constructed with nylon and polyester.

De Mestral named the product Velcro, a portmanteau of French words: *velour* (velvet) and *crochet* (hook). Today sixty million yards a year are sold through a multimillion-dollar company.

And that flashlight velcro'd to the wall by the back door will be just where you left it two years ago.

Sept. 13, 1955      G. DE MESTRAL      2,717,437

VELVET TYPE FABRIC AND METHOD OF PRODUCING SAME

Filed Oct. 15, 1952

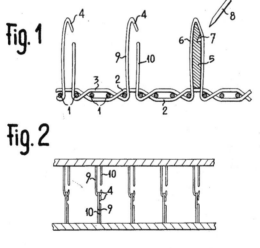

*INVENTOR*

*George de Mestral.*

*BY*

*ATTORNEY*

---

### Secret Communication System

US2292387A
Inventor: *Hedy Lamarr (et al.)*
Patent Granted: *August 11, 1942*

---

"The brains of people are more interesting than the looks I think," said actress Hedy Lamarr, billed as the Most Beautiful Woman in the World.

A world-class inventor with a far-ranging and facile mind, Lamarr was responsible for patenting a secret communication technology to aid the Allies in World War II. Quite an accomplishment for anyone and a source of great pride for Lamarr. But her resume had another notable entry: She had a glittering career as a successful Hollywood actress with starring roles to her credit alongside stars like Clark Gable and Jimmy Stewart!

Lamarr was born in Vienna and, as an ingenue, acted in a number of Austrian, German, and Czech films in her early film career. In 1937, with the winds of war gathering, she fled from her then-husband, a wealthy arms manufacturer, and secretly moved first to Paris and then London. There she met and impressed Louis B. Mayer, head of MGM, who was scouting for talent in Europe. He quickly secured her services with a $500-a-week contract and then began promoting her as the "world's most beautiful woman."

But Hedy had a bigger agenda than the "big screen." During World War II, the always curious Lamarr learned that radio-controlled torpedoes, an emerging technology in naval warfare, could be jammed and set off course. Her big idea? What if the incoming signal constantly shifted, a sort of frequency-hopping concept that could not be tracked or jammed? So, with her friend, composer and pianist George Antheil, Lamarr developed a device for synchronizing the signal with a player piano's internal mechanism. Together they drafted designs for the frequency-hopping system, which they patented in 1942 (her legal name, Hedy Kiesler Markey, is reflected on the patent application below).

Later, aviation tycoon and consummate ladies' man Howard Hughes, whom Lamarr briefly dated, was so captivated by her inventiveness that he put his team of scientists and engineers at her disposal, instructing them to do or make anything she asked for.

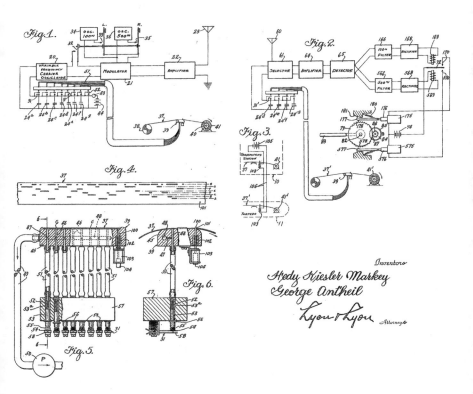

## Solar Panel

US389124
Inventor: *Edward Weston*
Patent Granted: *September 4, 1888*

There are a few things in life we can absolutely count on. One is that the sun will rise in the morning. This comfort and expectation so influenced humanity over the millennia that we saw fit to elevate the fiery orb beyond its physical presence and critical importance to that of a sky deity responsible for our happiness and welfare. In those early days before recorded history, sun worship had a wholly different connotation than lying on the beach in Miami, holding an umbrella drink.

Eventually, our curiosity being what it is, someone had to ask themselves the question: How can we harness all that power and put it to use?

In the early nineteenth century, a French physicist named Edmond Becquerel came to discover that certain materials, when exposed to light, generate a small electric current, now known as the photovoltaic effect. In 1839, he created the first photovoltaic cell by connecting silver chloride, placed in an acid solution, to platinum electrodes.

Just a few years later, in 1888, inventor Edward Weston received two patents for solar cells. His big idea, according to the abstract, was "to transform radiant energy derived from the sun into electrical energy"—the exact opposite of the way an incandescent light bulb works, converting electricity to heat, which then generates light.

Solar panel technology has continued to improve over the decades, and today solar panels are made primarily of silicon. Currently, the largest solar power plant in the world, India's Kamuthi Solar Power Project, covers about 3.9 square miles and has a power capacity of almost 650 megawatts. That's enough to power 422,000 homes—presuming that the sun rises in the morning.

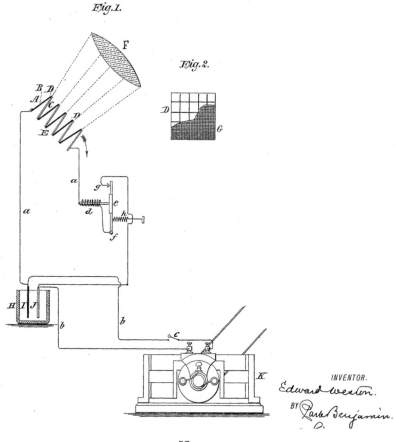

## Chicken Goggles

US730918A
Inventor: *Andrew Jackson Jr.*
Patent Granted: *June 16, 1903*

In 1903, Andrew Jackson Jr. had had enough. His chickens were in distress. They had begun to peck at each other's eyes for no discernable reason. Given that chicken social behavior was at that time—and probably still is—an area of uncertainty, myriad opinions, suggestions, and advice were volunteered from hither and yon: "They always seem to peck at spots—I had me feet pecked by free-range hens once because I had shoes with a white dot on the side. Maybe if you have something with a dot or two painted on, it would peck that instead." And "Chickens are prone to peck at things when they get bored. Put a mirror in, they will peck their own eyes then."

Jackson did know that chickens are instinctively cannibalistic and that pecking is the chicken's way of establishing hierarchy within the flock, thus, the term *pecking order.*

He also knew that there was no cognitive therapy that would convince chickens to stop pecking on their own, short of—you know—BBQ. So, he decided to take the decision out of their hands . . . er, talons, er, feet? Whatever.

He fashioned a set of goggles that were adjustable enough to fit any chicken, could easily be put on or taken off, and were relatively inexpensive to manufacture. And guess what? They sold pretty darn well. As a matter of fact, you can still get a pair of rather chic poultry specs today.

In 1955, Sam Nadler of the National Farm Equipment Company of Brooklyn appeared on *What's My Line?*, a popular prime-time television game show in which a panel attempted to determine the occupation of the contestant. Mr. Nadler's vocation was secretly revealed to the audience. The panel failed to guess his profession and later, Frank Heller, the show's director, said that the show's most unusual occupation over its entire run was ". . . the gentleman who makes eyeglasses for chickens."

No. 730,918.                                        PATENTED JUNE 16, 1903.

A. JACKSON, Jr.
EYE PROTECTOR FOR CHICKENS.
APPLICATION FILED DEC. 10, 1902.

NO MODEL.

## Electric Vote Recorder

US90646
Inventor: *Thomas Edison*
Patent Granted: *June 1, 1869*

This presidential election was fraught with discord and divisiveness over how votes were to be collected and counted. No, not the 2020 election. That, too, but we're thinking of the one in 1868!

That referendum, held on November 3, was the first election in the aftermath of the Civil War. Thanks to the First Reconstruction Act, it was also the first election in which African-Americans could vote. In the race for president of the United States, Ulysses S. Grant, the commanding general of the victorious Union Army and the Republican nominee, was up against the Democratic candidate, Horatio Seymour.

As history would later prove, Thomas Edison was ahead of his time—and his first patent a little too progressive. "The object of my invention," he wrote in his application for this patent, "is to produce an apparatus which records and registers in an instant, and with great accuracy, the votes of legislative bodies, thus avoiding loss of valuable time consumed in counting and registering the votes and names, as done in the usual manner."

How Edison's invention worked—or that Grant won the presidency—seems less important today than the effect the apparatus had on Congress. According to the intellectual property firm Suiter Swantz, which posts classic patents on its website, the members wanted nothing to do with a device that would speed roll-call voting, the method of the day. The slower the process, the more time they had to filibuster legislation and attempt to convince others to change their vote.

The Electric Vote Recorder was voted down, but its inventor was undeterred. Just twenty-two at the time, Edison would file another 1,092 patents to prove the naysayers wrong!

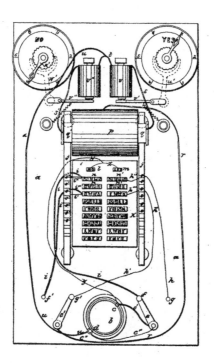

Witnesses.

Inventor.

## CHAPTER THREE

—

# Transportation and Space

The Starship *Enterprise* had a clear manifesto: "To boldly go where no man has gone before." And though its epic adventures took place in a distant future, its mission was a familiar one, and a theme that has echoed through history. Man is an explorer; it's in his DNA. This quest for new frontiers—above the clouds, beneath the oceans, across the continents—has driven many of humankind's greatest achievements. Orville and Wilbur Wright knew on that windy beach in Kitty Hawk, North Carolina, they were about to change history. Igor Sikorsky's helicopter had more in common with a hummingbird than a conventional aircraft. And John DeLorean's eponymous stainless-steel car, though it never really took off, was wildly imaginative, inspiring Dr. Emmett Brown's time-traveling vehicle in *Back to the Future*. As explorer Edmund Hillary said after conquering Mount Everest, "People do not decide to become extraordinary, they decide to accomplish extraordinary things."

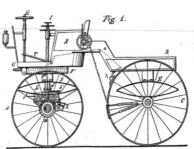

How to invent a car? Start with an engine and wheels. The rest—hood, roof, floor, doors, drink holder, CD player, etc.—will follow.

## Wright Brothers Flying Machine

US821393A
Inventors: *Orville Wright, Wilbur Wright*
Patent Granted: *May 22, 1906*

Humanity has always been jealous of the birds. For millennia, this longing to fly free, to shed the relentless clutch of gravity and take to the sky, has been an unending quest both in fable and reality. The Greek myth of Daedalus and Icarus described a set of wings fashioned of bird feathers and wax by Icarus's father as a means for his son to escape from a tower prison. His one warning to Icarus was not to fly too close to the sun for fear that the wax would melt. Of course, like any other rebellious kid, he ignored Dad's advice and ended up wingless in the briny.

The history of aviation extends from early forms of kite flying in China several hundred years BCE, through lighter-than-air balloons and on to the first primitive gliders. Of course, Leonardo da Vinci in the fifteenth century had some ideas about gliders and human-powered flight. His drawings are stunning in their prescience but were impractical to actually build, due to the lack of lightweight materials at the time.

It would take a couple of technically brilliant, ambitious bicycle mechanics in Dayton, Ohio to get the idea of powered flight off the ground. Between 1899 and 1905, the Wright brothers carried on a meticulous program of aeronautical research and bold experimentation that led to the first successful powered airplane in 1903 and a more sophisticated *Wright Flyer* two years later. It is thought by many that the brothers' big contribution to manned flight was a system of controls that allowed the pilot to precisely adjust the airplane's flight characteristics in real time.

When the Wrights finally took to the air on a beach near Kitty Hawk, North Carolina, on December 17, 1903, they made two flights each from level ground into a freezing headwind, gusting to 27 miles per hour. The first flight by Orville, who won the coin toss, was at 10:35 a.m., and traveled 120 feet in 12 seconds, at a speed of 6.8 miles per hour.

Today, a New York–to–London flight can be made in 4 hours and 56 minutes at a speed of 825 miles per hour, but you'll spend more time than that getting to and from the airport.

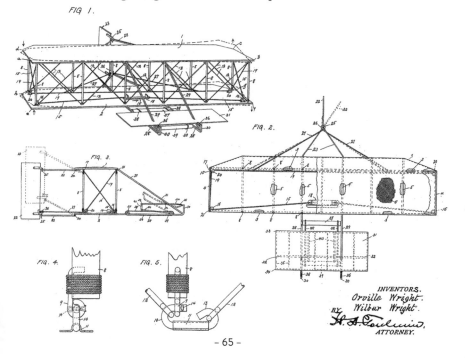

## Helicopter

USD150263 S
Inventor: *Igor Sikorsky*
Patent Granted: *March 19, 1935*

Italy in 1489. Observing the spinning action of a maple tree seed as it wafted gently to the ground, a man had an idea. What if the little winged whirligigs had their own source of power? Could their spinning motion counteract the force of gravity and thus cause them to rise?

At that time, this would have been impossible, beyond the scope of almost anyone on earth. But not for Leonardo da Vinci. In yet another stunningly prescient exercise, da Vinci put his thoughts on paper and designed a human-powered machine which, although impractically heavy, addressed the primary concept of a rotating-wing aircraft.

Fast-forward five hundred years: Igor Sikorsky, born in Kyiv Ukraine, was a bright, ambitious young man studying engineering in Paris and St. Petersburg. His efforts at the time culminated in a large, four-engine plane, completed in 1913, that so impressed Czar Nicholas II, the czar presented the young engineer with a diamond-studded gold watch for his achievements. Though designed as a civilian aircraft to carry passengers, the military converted it to use as a bomber during World War I.

After immigrating to the United States in 1919, Sikorsky founded the Sikorsky Aircraft Corporation in 1923 and developed the first of Pan American Airlines' legendary, ocean-crossing flying boats in the 1930s.

In 1939, Sikorsky designed and flew the VS 300, the first viable American helicopter, which pioneered the rotor configuration used by most helicopters today. The inventor soon modified the design into the Sikorsky R4, which became the world's first mass-produced helicopter in 1942.

During the Vietnam War, dense jungle growth prohibited the use of fixed-wing aircraft. But Sikorsky "choppers," also known as helicopter gunships, with their ability to land and take off vertically and rapidly in the most dangerous and precarious battle scenarios, proved themselves invaluable to American forces on both attack and rescue missions.

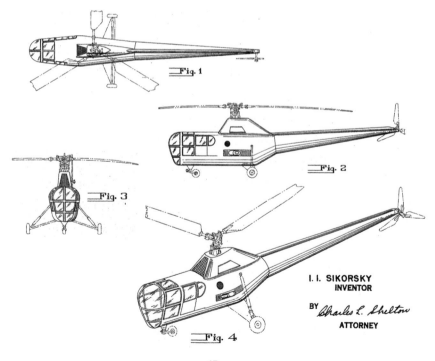

Fig. 1

Fig. 2

Fig. 3

Fig. 4

I. I. SIKORSKY
INVENTOR

BY *Charles L. Shelton*
ATTORNEY

## Submarine

US273851A
Inventor: *Jesse Jopling*
Patent Granted: *March 13, 1883*

It's easy to see why we have always been envious of the birds. What joy it would be to simply spread one's arms and take to the wild blue yonder! But the desire to swim as a fish in the cold, inky depths takes both a different mind-set and a rigid requirement—air.

This idea of underwater navigation has its roots in antiquity. Images of men using hollow sticks to breathe underwater have been found at the ancient temples at Thebes. At the siege of Tyre in 322 BCE, according to Aristotle, Alexander the Great used divers and conducted reconnaissance using a primitive diving bell. We imagine a rather short line of volunteers for that gig.

Even the great Leonardo da Vinci got in on the action. On one of the sheets of his *Codex Leicester*, da Vinci analyzes fish swimming in order to obtain information that might help humans stay underwater for long periods. However, he does not reveal it "because of the evil nature of men," as they would surely use it to sink enemy ships, causing the death of their occupants. Once again, the OG renaissance man proves that his knowledge of human nature is equal to his genius in countless other fields of endeavor.

Let's talk war. The first American military submarine was *Turtle* in 1776, a hand-powered, egg-shaped device, designed by the American David Bushnell to accommodate a single man. It's mission: to attack British vessels forming a blockade of New York harbor. There are no British records of an attack by a submarine during the war. Some have posited that the entire story was simply part of a disinformation campaign, designed to boost morale among the patriots, and that if an attack actually had occurred, it was from a rowboat.

It was simply "fake news," Revolutionary War–style.

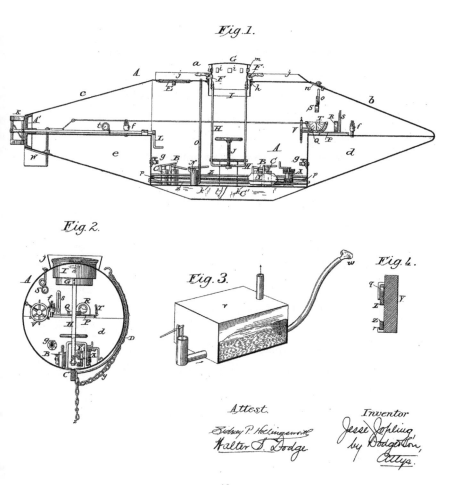

## Road Engine (first motorcar patent)

US549160
Inventor: *George Baldwin Selden*
Patent Granted: *November 5, 1895*

Ask almost anybody who invented the automobile, and odds are they'll say Henry Ford. But they'd be wrong. Ford's Model T was responsible for putting America, and later the world, on wheels. But many think that Ford's really big idea was his assembly line, which, in 1913, reduced the time it took to build a car from more than twelve hours to two hours and thirty minutes. But he wasn't the first. Consider German inventor Karl Benz, who patented his Benz Patent-Motorwagen in Germany in 1886, a date generally regarded as the birth year of the modern car. But tap on the brakes; there's a dark horse in the mix.

George Baldwin Selden, a patent lawyer and inventor, was granted a US patent for an automobile in 1895. Son of Judge Henry R. Selden, George, after several fits and starts, managed to graduate from Yale, pass the New York bar exam in 1871, and join his father's practice, all the while inventing a typewriter and a hoop-making machine in his basement . . . as a hobby.

Inspired by the internal combustion engine, designed by George Brayton, on display at the Centennial Exposition in Philadelphia in 1876, Selden commenced work on a smaller, much lighter version, succeeding in 1878, some eight years before the public introduction of the Benz Patent-Motorwagen in Europe. He filed for a patent on May 8, 1879. His application included not only the engine but its use in a four-wheeled car!

He then filed a series of amendments to his application, which stretched out the legal process, resulting in a delay of sixteen years before the patent was granted on November 5, 1895. He founded his own car company in Rochester; however, Henry Ford and four other carmakers took him to court. The legal fight lasted eight years, generating fourteen thousand pages of legal documents and a quote from Ford's testimony: "It is perfectly safe to say that George Selden has never advanced the automobile industry in a single particular . . . and it would perhaps be further advanced than it is now if he had never been born."

Tough cookie that Henry Ford.

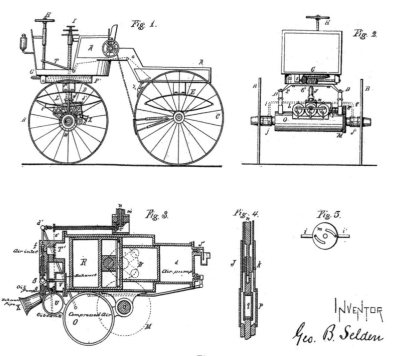

## Airplane Passenger Stacking System

US20150274298A1
Applicant: *Airbus*
Patent Filed: *March 26, 2015*
*Patent Pending*

If a patent filed in 2015 by Airbus Corporation ever gets granted, airplane passengers may one day be prone to complain even more about their fellow passengers—but this time from a whole new perspective.

In filing the application in March of that year, the giant aerospace manufacturer outlined the horrors of cramped seating. As described in the abstract accompanying the application, the company wrote, "In modern means of transport, in particular in aircraft, it is very important from an economic point of view to make optimum use of the available space in a passenger cabin. Passenger cabins are therefore fitted with as many rows of passenger seats as possible, which are positioned with as little space between them as possible."

What is also possible, Airbus reasoned, is an elevated deck of a wide-body aircraft to provide "a mezzanine seating area in a substantially unused upper lobe of the aircraft fuselage." In other words, to cram even more people into an airplane while giving them more legroom, essentially stacking passengers on top of one another.

But what about the lost baggage compartment above your head, you may ask. Not a problem: It's possible that, for a fee, you may be able to store your carrion (er, carry-on) on the midsection of the traveler above you.

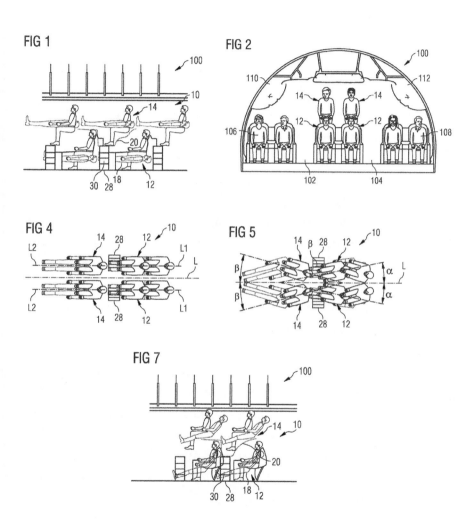

FIG 1

FIG 2

FIG 4

FIG 5

FIG 7

## ReAnima: Bringing the Dead Back to Life

US2014/0030244
Inventor: *Sergei Paylian*
Assignee: *Bioquark Inc.*
Patent Granted: *October 31, 2017*

How often have we said, "If only he were alive today!" or "How I wish she were still here with us"? Okay, so it depends on the deceased and the relationship we had with him or her. But, in general, alive is better than dead and any idea for bringing the dead back to life is worth at least a patent.

Humans have been trying to bring the dead back to life since time, well, in memoriam. Greek mythology preserved the quest in the story of Orpheus, a poet and musician who travels to the underworld to retrieve his recently deceased wife Eurydice. As Michael Sainato, a writer who has covered the subject, points out, the human death rate is 100 percent, yet that hasn't stopped the living from trying to find ways to reverse it.

More recently, in 2005, US patent #20050027316A1 was granted to Daniel Izzo for A Resurrection Burial Tomb, and in 2012, worldwide patent #WO2012090213A2 was granted to Pramod Kumar Nanda for the Resurrection Machine, which involved keeping dead bodies in a living chemical solution. Now comes Bioquark's ReAnima.

We're not really sure how this works; the patent application copy describes the process as follows: ". . . preparing a composition containing extracts of activated amphibian oocytes, the method where the composition is a pharmaceutical composition comprising an equal volume of the extra-oocyte composition and the intra-oocyte composition."

Still, if the process can somehow make our relatives who were dull in life more animated the second time around, we're all for it.

FIGURE 2

FIGURE 8

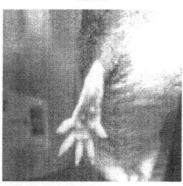

FIGURE 17

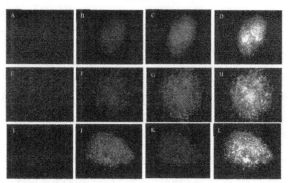

[0040] FIG. 2 is a photograph of a fully-developed mouse foot pad melanoma three weeks after inoculation with B 16 cells.

[0041] FIGS. 3, 4, 5, 6, 7, and 8 are photographs of a fully-developed (40 day postinoculation) mouse foot pad melanoma after 0, 10, 20, 35, 40, and 45 days treatment respectively, with the pharmaceutical composition of the described invention.

[0051] FIG. 17 is a series of photographs that show the expression of pluripotency markers by cells derived from human bone marrow stromal cells on d7 following co-electroporation with *Xenopus laevis* oocytes. (A)-(D) same field; (A) DAPI; (B) Oct 3/4; (C), Sox-2; (D), DAPI, Oct 3/4, and Sox-2 combined; (E)-(H) same field; (E) DAPI; (F) Oct 3/4; (G) Nanog; (H) DAPI, Oct 3/4, and Sox-2 combined; (I)-(I), same field; (I), DAPI; (J) Rex-1; (K) SSEA-1; (I) DAPI, Rex-1, and SSEA-1 combined.

## Human-Powered Flying Suit

US8087609
Inventor: *David A. Moore*
Patent Granted: *January 3, 2012*

Ever since our ancestors began falling out of trees, Homo sapiens have dreamed of flying. Some still do. David Moore, a former Air Force jet engine mechanic and founder of the website Aerosapiens, dreamed up this flying suit after studying, as the site notes, the flight of "dinosaurs, mammals (bats, flying squirrels), birds, reptiles, insects . . . as well as all archeological findings of man's ancient attempts at flight."

Perhaps "attempts" should have been emphasized.

Constructed of bright red nylon wings, pulled over a metal frame, and modeled on bats' arms and digits, Moore's contraption allows the pilot-operator to control direction and speed, and comes with a body harness to hold the wearer onto the suit. Good idea! Unfortunately, the Human-Powered Flying Suit's most successful flight lasted just 1.6 seconds.

Fortunately, however, this outfit isn't just for flying. According to the patent description, when they're tired of swimming in the air, pilots can manipulate the wings to walk on the ground. Someone interested in a career fighting crime could wear the outfit around Gotham, but not Moore. He reportedly had his sights set on a future Olympic Games, with athletes competing in extreme flying events clad in—you guessed it—Human-Powered Flying Suits!

Alas, that dream never got off the ground, either. Granted in 2012, Moore's patent has since expired.

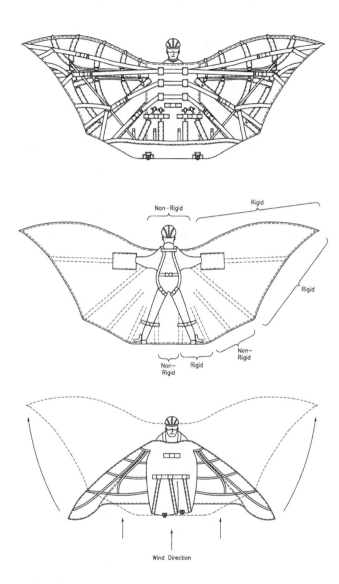

Non-Rigid

Rigid

Rigid

Non-Rigid

Non-Rigid

Rigid

Wind Direction

## Human Flight-Free Catapult

US5769724
Inventor: *Theodore F. Wiegel*
Patent Granted: *June 23, 1998*

Historians tell us that ancient invading armies used catapults to propel boulders, cannonballs, and even diseased bodies (or cows, if your knowledge of the medieval era comes by way of *Monty Python and the Holy Grail*) over castle walls, thereby alarming those on the other side. A thousand or so years later, one Theodore Wiegel thought it would be amusing to launch you and me into the air and possibly over socially restrictive walls.

As stated in his patent application, it is "an amusement ride for catapulting a human rider . . . into the air at a physiologically safe rate of acceleration along a predictable free-flight arc." (It's not specified, but exactly what a "safe rate of acceleration" might be when being propelled over someone's wall is something we'd like to know.)

As for the human catapult's potential uses, consider the possibilities: Gaining admittance to private clubs and gated communities from which we've been excluded; sending neighbors' stray cats (or cows) back over their fences; returning damaged packages in the general direction of the nearest Amazon warehouse . . . the sky's the limit!

In the end, of some concern to us is how the human catapult rides end. And here, the application spells out the details, more or less: "The rider is separated from the capsule and gently brought back to earth using an automatically deploying parachute or similar device."

While "brought back to earth" has a reassuring ring to it, we're wondering if Mr. Wiegel will be waiting on the other side to see how we are. His patent expired in 2016.

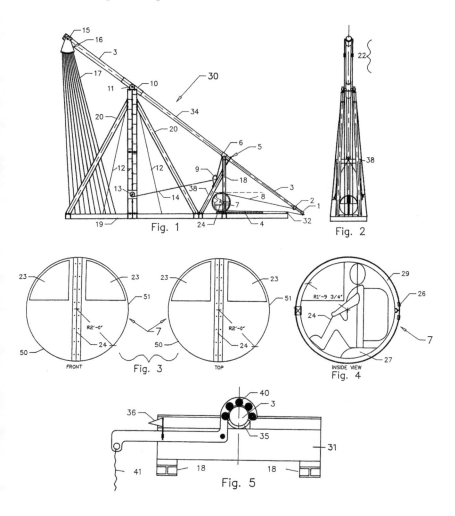

Fig. 1

Fig. 2

Fig. 3

Fig. 4

Fig. 5

## Satellite (satellite structure)

US2835548A
Inventor: Robert C. Baumann
Patent Granted: *May 20, 1958*

Embedded in every US patent application are the political and cultural undertones of the era in which it was filed.

In the fall of 1957, the United States and the former Soviet Union were locked in competition to develop military and scientific technology, including what would be the first intercontinental ballistic missiles and satellites. On October 4 of that year, the Soviets stunned America and the rest of the world when they successfully launched *Sputnik* 1, the first human-made object to orbit the earth. Two Soviet satellites—including *Sputnik* 2 with a Siberian husky named Laika onboard—quickly followed, ushering in the Space Age and the Cold War between the two countries.

Robert C. Baumann was a project manager at NASA (the US National Aeronautics and Space Administration) and satellite team leader when, on August 1, 1957, he filed the application for this patent covering the American response: *Vanguard* TV-3 (for Test Vehicle–Three). The three-pound aluminum sphere was little more than six inches in diameter.

Within moments after liftoff in December of '57, *Vanguard* I exploded on the launch pad, but it set the stage for the *Explorer* satellite, which was successfully launched into space on January 31, 1958. *Vanguard* II was launched on February 17, 1959, and, as of the summer of 2020, according to NASA, it remains the oldest human-made object in space.

Newspapers around the world reportedly had fun with the failed *Vanguard* mission, labeling it "Dudnik" and "Kaputik." Laika, *Sputnik 2*'s sole passenger, was probably wagging his tail.

## Apollo Spacesuit

US3751727A
Inventors: L. *Shepard, G. Durney, M. Case, R. Pulling, D. Rinehart, R. Bessette, A. Kenneway, R. Wise*
Patent Granted: *August 14, 1973*

"Always dress appropriately for the occasion" is a standard rule of etiquette. But when the occasion is arguably the greatest single achievement in human history, and when even a tiny design flaw could result in death, survival trumps fashion by a light-year.

So it was with the *Apollo* 11 spacesuits, technical marvels that had to be sufficiently flexible inside the lunar lander while as tough as suits of armor in the hostile environment of the moon.

In the pioneering race for space, time had entered the equation in the form of rigid completion deadlines. It sounds like lunacy, but the US Patent & Trademark Office took five years—from 1968 to 1973—to grant the patent for this otherworldly design. By then, the *Apollo* lunar landing was history. But the spacesuits Neil Armstrong and Buzz Aldrin wore on July 20, 1969 as they stepped onto the surface of the moon are as iconic today as the footprint the astronauts left in the lunar dust.

The *Apollo* 11 suits were designed at LLC Dover, a special engineering development and manufacturing company and a division of Playtex. Custom-made to fit each of the astronauts in the Apollo program, the twenty-one layers of gossamer-thin fabric, rubber, metal, and fiberglass were sewn to a tolerance of 1/64th of an inch by a team of eight women at the company's Dover, Delaware plant.

Materials used in the construction were themselves space-age at the time, including a water-cooled nylon undergarment. According to Howstuffworks.com, the suit and accompanying life-support backpack weighed 180 pounds on earth, but only 30 pounds on the moon.

A perfect marriage of technology and tailoring, they were flexible and tough enough to prevent punctures from micrometeorites and so sophisticated that Armstrong later remarked, "Those spacesuits were mini spacecraft."

The late '60s gave us some hip apparel, but none were as far-out as the *Apollo* 11 spacesuits.

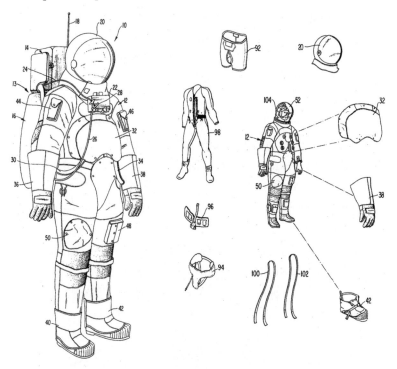

## GPS Tracking System (global positioning system)

US5379224A

Inventors: *Alison K. Brown, Mark A. Sturza*

Patent Granted: *January 3, 1995*

The question "Where am I?" has been around since the first hunter tried to find his way back to the safety of the campfire. Early humans had few constants in nature from which they could glean directional information, the most notable among them the daily arc of the sun.

Eventually, the stars divulged their awesome secrets and the magnetic compass brought the revelation of true north. Even finding one's bearings at sea, especially in daylight, would eventually be solved with the concepts of latitude, and—with accurate timekeeping—the big one, longitude.

But these astronomically based measurements didn't get the job done in the modern era, especially when up against the digital tsunami that came with the worldwide adoption of the computer. Precision and accuracy were now key for both civilian and military navigation.

Enter global positioning satellites, a system of geo-stationary satellites maintained under the auspices of the US government that remain in fixed positions relative to the earth's surface. Inventors Alison K. Brown and Mark A. Sturza patented a tracking system that would take advantage of the positional reliability of these devices to relay this critical information to receivers on earth. At no charge!

From the patent abstract:

"GPS is intended to be used in a wide variety of applications, including space, air, sea, and land object navigation, precise positioning, time transfer, attitude reference, surveying, etc. GPS will be used by a variety of civilian and military organizations all over the world."

Now, aside from its obvious strategic applications, GPS can tell a lost traveler where he is within six feet, anywhere in the world—or inform a worrying mother of her daughter's precise whereabouts after the prom.

Sorry, Jenny.

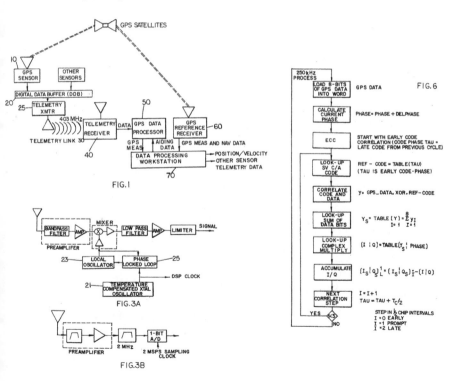

FIG.1

FIG.3A

FIG.3B

FIG.6

## Telescope

US8509A
Inventor: *Alvan Clark*
Patent Granted: *November 11, 1851*

Sixty-eight hundred years ago, small bands of Paleolithic hunter-gatherers gazed in wonder and awe as a cosmic visitor—a deity of ineffable brilliance and speed—blazed across the night sky.

What our ancestors observed was a comet—a celestial snowball of rock, frozen gases, and dust—whose orbit around the sun is so wide that it wasn't until March 7, 2020 that NASA scientists discovered it. They employed the telescope for which the object is named—the Near-Earth Object Wide-field Infrared Survey Explorer, or NEOWISE—which uses infrared sensors to detect objects too distant and cold to be observed in visible light. With that telescope they detected what only their Creator had seen before: to date, 158,000 asteroids and more than 155 comets.

And, if turned in the opposite direction, the telescope would have looked down the long trail of instruments that define humanity's quest to understand the universe.

The inventor of the modern telescope, and the applicant of the first telescope to be patented in the United States, was Alvan Clark of Cambridge, Massachusetts. In business with his two boys, Alvan Clark & Sons ground the lenses for the forty-inch refracting telescope at Yerkes Observatory in Wisconsin in 1897, and, to this day, it's the largest refracting telescope in the world.

Clark's patent simply covers an improvement on all the telescopes that came before him—from Hans Lippershey (see page 34), the Dutch lens maker who in 1608 applied for a patent for a refracting telescope, to the titans of astronomy: Galileo Galilei, Johannes Kepler, and Sir Isaac Newton, builder of the first reflecting telescope.

Sixty-eight hundred years from now, a being vaguely resembling us may be looking back at earth—through a much more powerful telescope from a very distant planet.

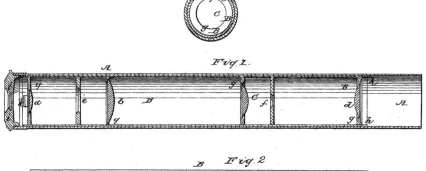

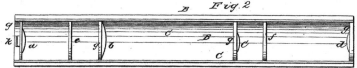

## Headlight for Horses

US911733A
Inventor: *P.A. Libby*
Patent Granted: *February 9, 1909*

In 1908, Henry Ford was about to change the course of world history with the introduction of the Model T, the first affordable automobile, at a price of $260.00. Eventually, fifteen million of Ford's horseless carriages would be sold. However, that year, the nation was still getting around either on or behind their trusty nags—in daylight.

Come nightfall, all bets were off. Although equine vision is superior to humans' in some aspects, the pitch black of the countryside was well beyond the limits of a horse's visual acuity, making a midnight gallop through the woods not just foolhardy but suicidal. It became clear that some light needed to be shed upon this problem.

For untold centuries the oil-soaked burning torch, occasionally accompanied by a pitchfork, had been the preferred on-horseback lighting device for unruly mobs, drunken husbands, and suffragists. But in 1908, after failed attempts by others to strap a burning lantern to the top of a horse's head, a certain Eugene Richards, a man who clearly had things to do after dark, patented a Horse Headlight, powered by an early version of a battery and cinched around the horse's neck, facing forward.

This made perfect sense but for one small detail. Horses are prey animals and their first defense against a threat is to run. So, when the horse's natural movement caused the light beam to bounce erratically in front of him, the animal's instinct was to bolt immediately for

Belgium or some other unplanned destination, conceivably giving rise to the phrase "Whoa, Nelly!"

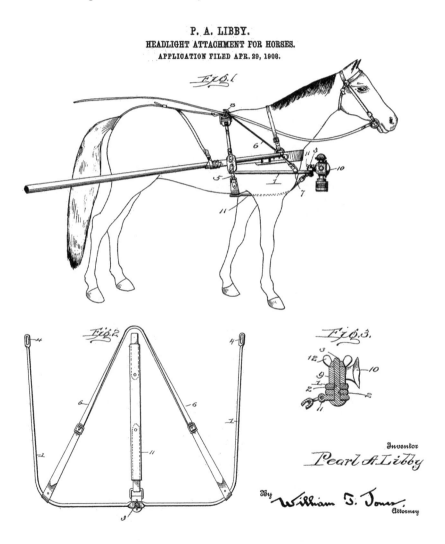

P. A. LIBBY.
HEADLIGHT ATTACHMENT FOR HORSES.
APPLICATION FILED APR. 29, 1908.

Inventor
Pearl A Libby

By William T. Jones,
Attorney

## DeLorean Automobile

USD283882S
Inventor: *Giorgetto Giugiaro*
Patent Granted: *May 20, 1986*

John Z. DeLorean, it can be said, invented the muscle car, and forever changed the way America related to the automobile. But it was a star turn by the eponymous automobile as a time-travel device in *Back to the Future* that sealed his place in popular culture.

DeLorean was a brilliant engineer and a fearless visionary, but, most of all, a talented maverick, infamous for bucking the conservative corporate status quo during his tenure as the youngest division head in General Motors' history.

His reputation was made when he recognized a pent-up demand among younger car buyers for more power and performance. His big idea was a simple one: Take an existing platform, in this case the 1964 Pontiac Tempest/Le Mans, offer an option package with a larger, more powerful engine, and rebrand it as the GTO after Ferrari's legendary 250 GTO sports car. It would become one of the most coveted American performance cars of all time.

But all this success wasn't enough to feed the glittering ambitions of DeLorean, who broke away to start the DeLorean Motor Company (DMC) in 1973 and bring to market the DeLorean, a futuristic concept car, based on drawings by famed Italian designer Giorgetto Giugiaro that featured gullwing doors and a body of stainless steel. Unfortunately, it was doomed by production delays and never reached the consumer until 1981, when it was greeted with lukewarm reviews from both critics and the public.

And so began one of the most shocking and dramatic corporate downslides of the modern era. In October 1982, with DMC insolvent and in debt for $17 million, DeLorean, in desperation, entered a scheme with an FBI informant to sell 220 pounds of cocaine, worth $24 million. DeLorean, was able to successfully defend himself at trial, using the procedural defense of police entrapment. The trial ended in a not guilty verdict in August 1984, by which time DMC had declared bankruptcy and shut down.

But DeLorean, the man, the machine, and the "flux capacitor," live on in the space-time continuum.

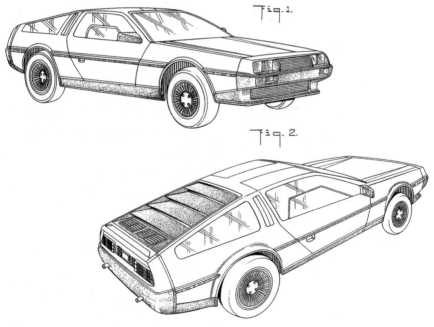

Fig. 1.

Fig. 2.

## Parachute

US1108484A
Inventor: *Štefan Banič*
Patent Granted: *August 25, 1914*

In concept, the principal use of a parachute has always been to mitigate the force of gravity by using air resistance to safely slow the descent of a human plummeting to earth. As such, the familiar images associated with the parachute often depict thousands of airborne infantrymen floating down from the sky, like so many dandelion heads during the D-Day invasion, or today's colorful, high-tech chutes that can be steered with pinpoint accuracy, gliding for miles to gently land on a target the size of a ping-pong table.

The seminal idea for the parachute, like so many other prescient concepts, is thought to have originated in the fertile mind of Leonardo da Vinci, this one in 1483. But it was in 1797 that a Frenchman named André-Jacques Garnerin ascended in a hydrogen balloon to an altitude of 3,200 feet, severed his connection to the mothership, and rode his frameless silk parachute to the ground. He miraculously landed shaken but alive.

An early attempt at modern parachute design was the brainstorm of Štefan Banič, a Slovakian coal miner living in Pennsylvania. After witnessing a plane crash in 1912, the inventor constructed a prototype and was granted a US patent in 1914. The aerodynamically oddball design was radically different, a sort of umbrella attached to the body. It was later claimed, to some skepticism, that he successfully tested it in Washington, DC by jumping first from a fifteen-story building and later from an airplane in 1914. He donated his patent to the US Army. There is no evidence that it was ever used.

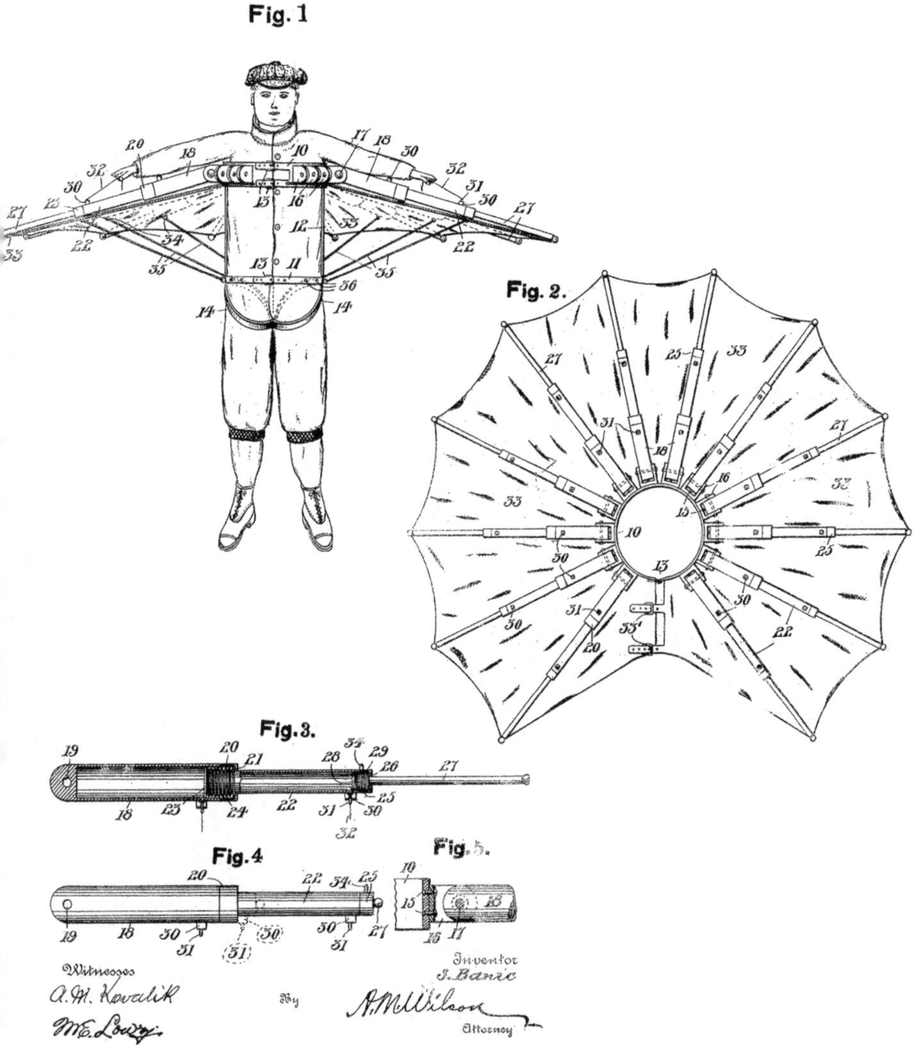

**Fig. 1**

**Fig. 2.**

**Fig. 3.**

**Fig. 4**

**Fig. 5.**

Witnesses
A. M. Kovalik
M. E. Lowrj.

Inventor
S. Banic

By
A. M. Wilson

Attorney

## Electronic Landing Marker

US10395544B1
Inventor: *Scott Raymond Harris (et al.)*
Patent Granted: *August 27, 2019*

We are spoiled by the conveniences of modern life. Why endure the painful, physical stress of holding a book in your hand when thousands of them will fit on an iPad? Why consult a map to find your way to a destination when the talking GPS on your cell phone will safely (most of the time) guide you to within six feet of your target? And why get out of your pajamas, hike all the way to your car in the garage, and suffer an endless air-conditioned drive to the mall, when you can, with a couple of taps on your laptop, select, pay for, and expect two-day delivery of your item to almost anywhere on the globe—with obvious exceptions like a secret village in Borneo or Mount Everest Base Camp.

Why indeed? That's a question being asked these days at Amazon, the world's largest retailer. And they think they've got a better idea. It involves drones.

It seems like mere minutes ago that the drone was a newfangled contraption one could drive around in the air using a cute little radio-controlled joystick. Then the drone grew up right under our noses. It got more sophisticated, powerful, and capable of carrying bigger loads for longer distances—loads about the size of an Amazon package.

Okay, that sounds fine, but even a non-nerd knows the next question: How does it know where to land? Amazon's answer: on a specialized mini–landing pad that signals the approaching drone carrying your twelve-pack of razors and guides it in for a landing. Sounds great, right? Have you ever heard a drone? Even a small one sounds like an amplified swarm of bees. A drone capable of toting your family-size box of laundry detergent would be loud enough to invite every dog in the neighborhood to a special event in your front yard.

And cats? By the time that thing lands, they'll be in the next state.

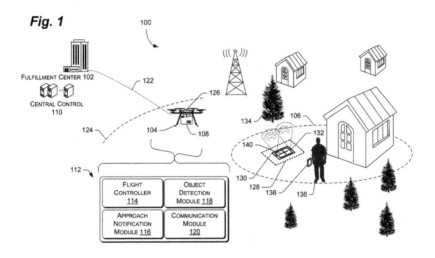

**Fig. 1**

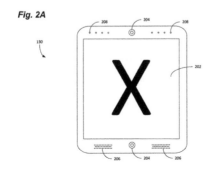

**Fig. 2A**

## Zeppelin (navigable balloon)

US621195A

Inventor: *Ferdinand Adolf Heinrich August Graf von Zeppelin*

Patent Granted: *March 14, 1899*

Fasten your seat belts, ladies and gentlemen (*Schnallen sie sich an, meine Damen und Herren*) and come immediately to attention. Introductions are in order:

Ferdinand Adolf Heinrich August Graf von Zeppelin was born on July 8, 1838, the scion of the German noble family Zeppelin from the community of Zeppelin outside the town of Butzow in Mecklenberg. Ferdinand was the son of Württemberg Minister and Hofmarschall Friedrich Jerôme Wilhelm Karl Graf von Zeppelin and his wife Amélie Françoise Pauline. He attended the military academy at Ludwigsburg near Stuttgart, and at age twenty became an officer in the army of Württemberg.

Now that you're clear on the family history, let's take off.

In 1863, with a pass signed by President Abraham Lincoln, the count traveled to the United States as a military observer during the American Civil War. But it was in St. Paul, Minnesota—far from the battlefields—that Zeppelin saw and rode in his first lighter-than-air balloon. Floating seven hundred feet high in a tethered ascent, Count von Zeppelin had seen the world from the air, giving rise to his grand idea, a navigable balloon. He began developing preliminary concepts for the design of a steerable airship. But it was not until his retirement from the Army in 1890, at the age of fifty-two, that Zeppelin was able to devote himself more fully to the problems of lighter-than-air flight. Within ten years he would build his first airship, *Luftschiff Zeppelin* 1 (LZ-1).

It wouldn't be until after his death in 1917 that his namesake, the *Graf Zeppelin*, made its very first commercial passenger flight across the Atlantic, departing from Friedrichshafen at 7:54 a.m. on October 11, 1928, and landing at Lakehurst, New Jersey, on October 15, 1928, after a flight of 111 hours and 44 minutes.

It is unclear whether peanuts were served on the flight.

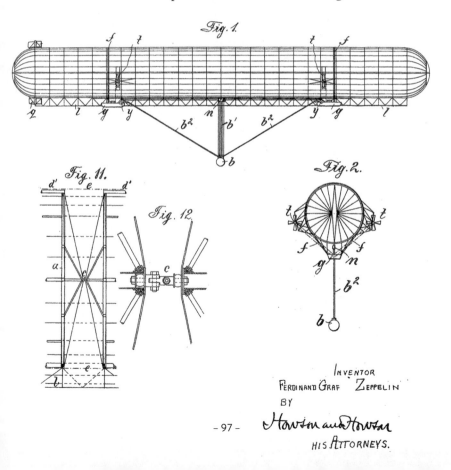

## Windshield Wiper System with Intermittent Operation

US3351836A

Inventor: *Robert W. Kearns*

Patent Granted: *November 17, 1967*

It's one of life's little mechanical conveniences that we rarely think about, but use all the time. Usually found just past the on-off switch for the windshield wipers on your steering wheel, it's the intermittent wiper speed adjustment. Who among us hasn't played the little game of perfectly adjusting the speed of the wiper to match the intensity of the rain? It's one of those gadgets that just plain works. The story of its inventor and the epic patent battles he fought with Detroit auto manufacturers is right out of a Hollywood movie. But that would come later.

It all began with the pop of an errant champagne cork on engineer Robert Kearns's wedding night in 1953. It hit him in his left eye and left him legally blind for life. Nearly a decade later, while driving through a light rain, the constant movement of his car's wiper blades irritated his already troubled vision. That's when he got the idea. What if the wipers could act more like the human eye, which blinks intermittently?

He presented his idea to Ford representatives who were enthusiastic about the concept at first, but later appeared to have lost interest. That is, until they introduced the feature on their cars in 1969. Kearns, acting as his own lawyer, challenged the automaker and refused offers of a settlement. Eventually, after a titanic struggle, he won.

Kearns's legal battle against Ford became the basis for a full-length biographical feature film, *Flash of Genius*, in 2008. As for the lawsuit, Kearns sought $395 million in damages, but was awarded $10.2 million—in one payment, not intermittently.

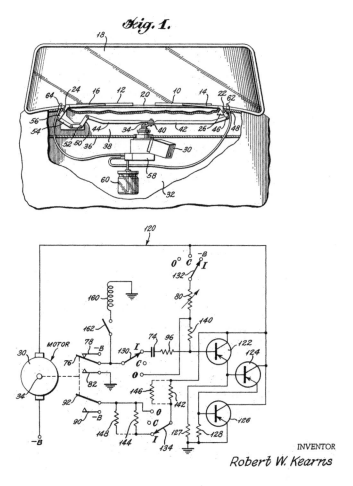

**Fig. 1.**

INVENTOR

*Robert W. Kearns*

BY

*Lane, Aitken, Dunner & Ziems*

ATTORNEYS

| Vision System for an Autonomous Vehicle (self-driving vehicle) |
| --- |

US20070291130A1
Inventor: *Alberto Broggi (et al.)*
Patent Granted: *June 22, 2006*

What do a tank, a taxi, a limousine, and an off-road 4x4 vehicle have in common? They all require a driver. And, of course, scientific progress being what it is, it's probable that the very day a motorized vehicle first rolled down the road, someone was trying to figure out a way to get it to drive itself.

Not an easy task. Imagine the variables that must be taken into account for a vehicle to drive autonomously. Is it light or dark outside? Where's the road? Indeed, what's a road? What's a traffic light, a street sign, a pedestrian, how fast should you go, where to stop? You get the idea.

It would take until the 1950s and the advent of the computer to get things rolling. The first semiautomated car was developed in 1977 by Japan's Tsukuba Mechanical Engineering Laboratory. It required specially marked streets, interpreted by two cameras on the vehicle, and an analog computer, which, by today's standards, had the processing power of that $5 calculator in your desk drawer.

By 2006, the world had changed. That little calculator now had the power to compute Pi out to a few hundred places or to process enough information to drive a car down the road. Alberto Broggi and his team patented a system that was configured to receive, by way of multiple video cameras, information on the surrounding terrain, upcoming obstacles, or a particular path, and to automatically respond just as a human operator would. Today, we're getting closer by the minute to a real-world, self-driving car, but there will never be a digital substitute for a kid in the back seat asking, for the fifty-third time, "Are we there yet?"

FIG. 1

FIG. 3

FIG. 5

FIG. 4

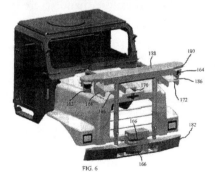

FIG. 6

## Life Raft

US258191A
Inventor: *Maria Beasley*
Patent Granted: *May 16, 1882*

Maria Beasley was born Maria Kenny in Philadelphia in 1847. Not much is known about her early life or education, and it remains unclear why she was inspired to become an inventor. But there is speculation that she was influenced by the Women's Pavilion at the Centennial Exhibition of 1876 in Philadelphia, the first official World's Fair, celebrating the signing of the Declaration of Independence.

She was married to Samuel Beasley in 1865, and ran the household in addition to her business and creative pursuits. Philadelphia directories from 1880 list her as a dressmaker. That moniker wouldn't last too long, however, as her creativity, talent, and ambition came into play.

At age forty-four, Maria put down her sewing needle and embarked on a somewhat unexpected new career as a serial inventor. Her first patent, granted in 1878, was for a barrel-hooping machine to greatly speed up the manufacture of barrels. Her invention produced 1,500 barrels a day and earned her a "small fortune," according to the *Evening Star* in 1889. Other inventions would include foot warmers, cooking pans, and anti-derailment devices for trains. But it was her improved life raft design that brought her international renown. Most rafts of the era were simply made with modest planks of wood, but Maria's design was collapsible, fireproof, compact, safe, and easy to launch.

These innovative life rafts entered history onboard the world's most infamous passenger liner, the *Titanic*, when it sank in the North Atlantic in April 1912. Of the over 2,000 people onboard, more than 1,000 people were lost. However, the ship carried twenty of Beasley's life rafts, which enabled 706 men, women, and children to survive safely until help arrived.

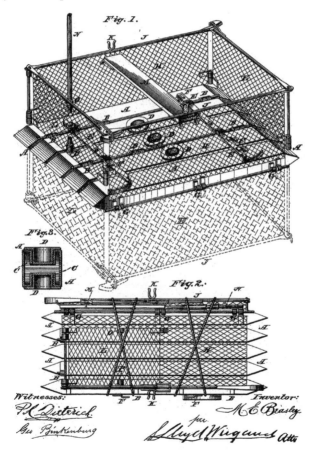

## CHAPTER FOUR

—

# Health and Medicine

Before many of the patents in this section were granted, life for millions of people was at best difficult and at worst wretched. The Cardiac Pacemaker miraculously restored rhythm to irregularly beating hearts. Insulin brought relief from agonizing pain to those afflicted with diabetes. A number of Brain Implant devices have restored entire regions of brain tissue impaired by illness or accident. Also to be found here, however, are cautionary tales, as in the case of the Blood Analysis Machine, a fraudulent device that rivaled the best of scams in a Mark Twain novel. Saving the day for us, meanwhile, was coming across some of the wackiest patents imaginable—an experience akin to opening the bathroom door at an opportune moment. We need only mention the Toilet Overspray Device and Forehead Support Device to get our point across, proving that the bathroom may be the ideal laboratory for the flatulent patent!

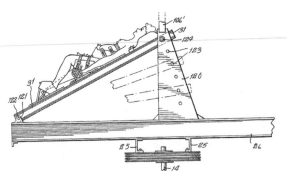

"Apparatus for Facilitating the Birth of a Child by Centrifugal Force,"
which gave birth to the concept of postnatal flight.

| Brain Implant Device |
| --- |
| EP1614443A1<br>Inventor: *Joseph H. Schulman*<br>Patent Granted: *September 19, 2007* |

For decades, scientists have romanced the marriage of brain and machine. Researchers at the University of California began experimenting with Brain-Machine Interface (BMI) and Brain-Computer Interface (BCI) in the 1970s. By the mid-1990s, after years of experimentation on lab animals, the first neuroprosthetic devices had been implanted in humans.

The technology surrounding brain or neural implants has evolved, in part, to circumvent regions of the brain damaged by birth defects, head injury, strokes, Parkinson's disease, or clinical depression.

Early experiments led to the development of cochlear implants that translate audio signals into electrical pulses sent directly to the brain of hearing-impaired patients, allowing the hearing impaired to hear. Later, baby aspirin–size computer chips, implanted in the motor cortex region of the brain, helped paraplegics and quadriplegics to control tablets simply by thinking about moving and clicking a cursor.

Schulman, who holds more than one hundred patents in biomedical technology, invented this implantable device to detect and transmit electrical signals to and from the brain. The signals are sent to a remote central processor that wirelessly controls body functions. (The signal Dr. Schulman sent our brains was to include his patent in this book.)

Now, scientists and entrepreneurs are exploring ways that Artificial Intelligence (AI) can enhance human-computer interfaces.

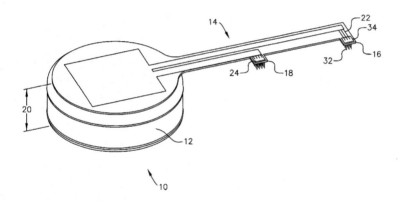

## Tampon (catamenial device)

US1926900
Inventor: *Earle C. Haas*
Patent Granted: *September 12, 1933*

It isn't every day that a man does something to make a woman's life better. In fact, in the history of relations between the sexes, it's safe to say the opposite has been true. Nevertheless, on September 12, 1933, Denver, Colorado physician Dr. Earle Haas was granted a patent for this product to ease his wife's monthly discomfort.

Women, naturally, hadn't been waiting around for a gentleman like Dr. Haas to address the problem. In ancient Rome, women reportedly fashioned tampons out of wool, and in Egypt and Japan, from papyrus and paper. (But don't rely on us; a more knowledgeable source is Nancy Friedman, whose book *Everything You Must Know about Tampons* appears to be a staple of tampon literature.)

Hospitals in Dr. Haas's day used compressed cotton pads, called tamponades, to absorb blood during surgery. Wanting to keep the pads sanitary for his wife and other women, Haas created the cardboard applicator with string, and patented what would become the modern-day tampon as a "catamenial" (Greek for "monthly") "device," and three years later sold the patent to the founder of Tampax.

As such, the good doctor has become a kind of menses-mentor to women today.

"Talk about sympathy pains!," Victoria Ainsworth wrote in a blog about his invention on the website The Spot. "It was Earle

Haas—doctor, businessman, humanitarian, husband-of-the-year—who invented the tampon we all know and love.

"Oh," Ainsworth adds, "and he invented the diaphragm, too. Pretty, pretty, pretty good."

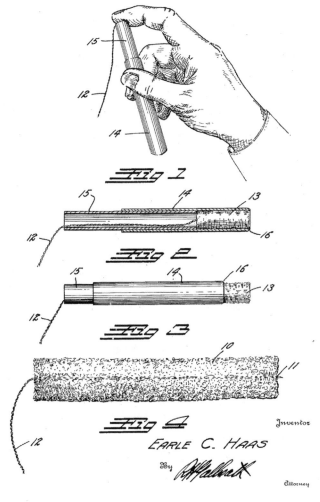

*Fig 1*

*Fig 2*

*Fig 3*

*Fig 4*

Inventor

EARLE C. HAAS

By

Attorney

## Viagra

US6469012B1
Inventors: *Peter Ellis, Nicholas Kenneth Terrett*
Patent Granted: *October 22, 2002*

It has attained legendary status as one of science's great achievements, especially by those who benefit from its life-enhancing effects. But the drug the world knows as Viagra (Sildenafil) did not begin as an erectile dysfunction pill. The compound was originally developed and patented by Pfizer as a drug called Revatio to treat high blood pressure and chest pain due to angina.

However, clinical trials revealed a rather spectacular side effect—in fact, it was more effective at inducing an erection than anything else. The drug was then renamed and patented as Viagra to the delight of millions of romantically challenged men the world over.

If ever there was a sure thing, it was Viagra, with over $1 billion in sales in the drug's first year of production. But, as it has since time immemorial, male virility proved to be a touchy area and numerous lawsuits were filed against Viagra and Pfizer. One suit for $110 million, on behalf of a car dealer from New Jersey, claimed that he drove his car into two parked vehicles after Viagra caused him to see blue lightning coming from his fingertips.

Over the years, there have been numerous attempts to create a Pink Viagra to boost women's libidos. However, all attempts have failed based on the unscientific belief that a woman's main sex organ is "between her ears, rather than between her thighs."

## Blood Analysis Machine

US9858660
Inventor: *Elizabeth A. Holmes*
Patent Granted: *February 15, 2013*

Elizabeth Holmes was always the smartest person in the room, or so she thought. Her spectacular rise and fall, from geek wunderkind college dropout, to the youngest, wealthiest self-made female billionaire in America, to defendant in a massive fraud case, have become the stuff of corporate legend.

Over the course of her stunning collapse, *Forbes* magazine revised its estimate of her net worth to zero and *Fortune* named Holmes one of the "World's Most Disappointing Leaders." And it all began because Elizabeth hated needles.

During her freshman year at Stanford, Holmes worked in a laboratory that tested for acute respiratory syndrome through the collection of multiple blood samples with syringes. There, she had her big idea to "democratize health care" by performing diagnostic blood tests quickly and with only a tiny amount of blood from a finger prick. Her professors told her it wouldn't work. But, as is the case with so many driven inventors, those warnings served only to steel her will to press on.

In 2003, Holmes formed Theranos, a portmanteau of "therapy" and "diagnosis."

Now it was time to unleash her natural talent as a public relations wizard. Always an admirer of Apple founder Steve Jobs, she deliberately copied his style, frequently appearing in a black turtleneck.

And so, by dint of her personal magnetism, facile salesmanship, and striking, unexpected baritone speaking voice, the investors scrambled onboard and the money poured in.

But there was a problem. The testing machines didn't work and she couldn't deliver on her promises. Then, cornered, she went off the ethical rails, commencing what could be construed as a high-tech version of a Ponzi scheme. Eventually, the legal system caught up with her and the jig was up. (The patent is currently in litigation.) At its height in 2015, Theranos had more than eight hundred employees. By 2018, fewer than two dozen remained and, eventually, most of them were laid off.

It was a "bloodless" coup.

## Male Chastity Device

US860779B1
Inventor: *Youxian Ma*
Patent Granted: *December 17, 2013*

We're all grown-ups here, so let's talk about it. When the first stone-agers sat around a campfire chewing on the burnt yak-of-the-day, the après-dinner conversation, such as it was, would surely have turned from tall tales of hunting prowess to even taller tales of sexual prowess. (It's possible that during this early social intercourse, the lie was invented . . .)

In those wooly days, the male urge for procreation was unconstrained, much to the dismay of the local ladies who had no vote whatsoever in the proceedings and relied on foot speed as their only prophylactic.

Today, millennia later, the talk around the BBQ remains basically the same, except for one major factor—women have the power.

That said, the male urge, though considerably chastened by evolution, revolution, religion, STDs, and marriage, still rears its ugly head at the most inauspicious times. Apparently, this problem was the seminal idea behind the invention of the Male Chastity Device, a draconian apparatus that served as a mechanical cease-and-desist order for unwanted male ardor.

Actually, just reading the patent application abstract should be sufficient to quell any impulses:

"The device comprises a first loop, a cup, a second loop hinged to the first loop and optionally to the cup, and a removable fixing mechanism. The first loop is configured for being traversed by the user's penis and scrotum, and for encircling the penis behind the scrotum. The second loop is configured for being traversed by at least the penis, and for encircling the penis ahead of the scrotum, so that the user's testicles are located between the first loop and second loop. The cup is configured for enclosing at least a head of the penis. The removable fixing mechanism is configured, when joined to the device, for preventing removal of the device from the penis by limiting or blocking relative rotation between the first loop and the second loop, and, if needed, by limiting or blocking relative rotation between the cup and the second loop."

We have a headache.

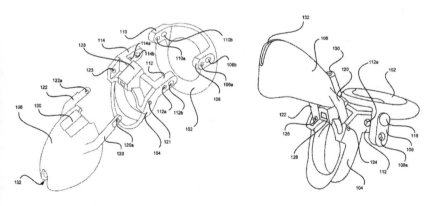

This device, like romance in general, was complicated.

## Apparatus and Method for Sharing User's Emotion

US20130144937A1
Inventor: *Hosub Lee*
Assignee: *Samsung Electronics Co., Ltd*
Patent Granted: *June 6, 2012*

Emotions used to be something we kept to ourselves or shared with loved ones and close friends. As social media has expanded our circle of friends and acquaintances, media and telecommunication companies have been pressing their faces against the glass for closer and closer looks into our relationships.

Now, we find, Samsung and other companies have been trying to look into our hearts!

According to IPWatchdog.com, an online intellectual property site that tracks patents, what this device from the South Korean electronics conglomerate sought to do was analyze social network users' posts and interactions, then rate any change in their emotional states. The apparent purpose was to improve communications by limiting interactions that produced negative emotions and by increasing those that produced positive emotions.

But then what—feel good or bad for us? Weep? We have our doubts. Something mercenary, if not nefarious, may have been in the works, such as selling the raw meat of our emotions like a commodity. Whatever the case, we didn't have to worry for long.

Samsung abandoned the patent for the device, yet, to us, its short-lived existence signaled a further intrusion into our lives and an exploitation of human instincts. Why the company lost interest we don't know, but if the inventors are feeling abandoned and want to share how they feel, we're here for them.

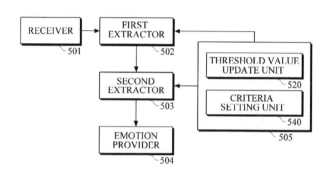

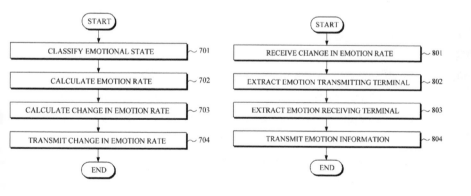

## Smart Football Helmet (systems and methods for helmet liner evaluation)

US20160320278A1
Inventors: *Bill Gates (and nineteen others)*
Patent Granted: *January 22, 2019*

Given the rise of brain injuries among young football players and professional players alike, concern over concussions among football players has been on the minds of many in recent years. When one of the minds is Bill Gates, concern can pivot into potential solution, and in a hurry! (Gates was reportedly drawn to the venture out of his relationship with childhood friend Paul Allen, cofounder with Gates of Microsoft, owner of the American football team the Seattle Seahawks.)

In 2014, the Microsoft cofounder and billionaire philanthropist, along with Amazon's Jeff Bezos and Jack Ma, cofounder of Alibaba Group, convened the seventeen other members of the nonprofit Intelligence Ventures in a room in Bellevue, Washington to tackle the problem (sorry, okay?) head-on. When they emerged eight hours later, according to published reports, the group had a design for a helmet and liner connected to internal sensors that assess the risk of concussion to players in real time.

Described as a kind of instant diagnostic test, when an impact to the helmet occurs, the embedded sensors determine the force, acceleration, and torque of the hit, and software assesses its severity and monitors thresholds for concussions, as well as head and neck injuries. All that data then gets communicated instantly to doctors and coaches on the sidelines.

The group applied for a patent for the helmet in 2014. The lesson here for would-be inventors and innovators? Even with friends in high and powerful places—Ma. Bezos. Allen. *Really?*—be prepared to wait on the sidelines a really long time for your patent application to be approved!

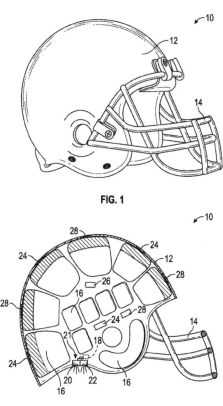

FIG. 1

FIG. 2

## Toilet Overspray Device for Males

US6357055B1
Inventors: *Eve Gambla, John Manzella*
Patent Granted: *March 19, 2002*

If there is one thing a male family member has on his mind when attempting to perform number one, it is the profound apprehension of missing the target or, worse, missing the toilet altogether. Although this tendency occurs naturally in three-year-olds and in seniors over seventy, it is also present in 98.7 percent of the male population.

Performance Anxiety in Urinators (PAU) syndrome is often due to the tension arising from the clear and present danger that any overspray will result in family strife, marital discord, and, in some extreme cases, violence and death. The Toilet Overspray Device was designed to optimize the target area, thus ensuring a direct hit in instances that do not involve consumption of alcoholic beverages.

In daily use, the device may be stored in any toolshed, root cellar, or Bass Boat. When needed, the user need only fetch, position, and calibrate the device within the toilet bowl. Side curtains provide additional protection and a central target—inspired by World War II's legendary Norden bombsight—offers a clear aiming point. Setup time is usually under twenty minutes and some hygiene is possible by hosing it down outside, then drying it in the sun or with a heat gun or a hair dryer.

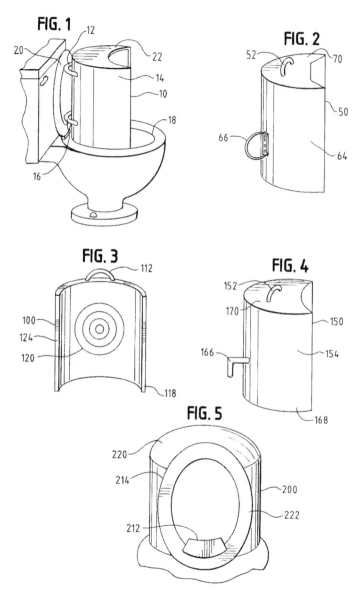

FIG. 1

FIG. 2

FIG. 3

FIG. 4

FIG. 5

## Forehead Support Device

US6681419B1
Inventor: *Eric D. Page*
Patent Granted: *March 5, 2002*

The items necessary to support human life comprise a short list: water, fire, food, and shelter from the elements. Beyond these basics and over the span of millennia, the list has grown longer and more complicated. Today it might include a lifetime HBO subscription, a beach house, and a rich uncle in his final fight against malaria.

However, there are times when a very specialized type of support becomes a necessity. So thought inventor Eric Page when he found himself standing erect in front of a male urinal, feeling a bit shaky and in need of some assistance. So, with (presumably?) both hands already in use, Page simply leaned forward and rested his forehead against the wall above the urinal, thus creating a third balance point and instantly calming his urinary trepidations. "Now, if only there was a comfortable place on which to rest my head, I'd have the problem firmly in hand," the inventor must have thought.

Soon, this little drip of an idea gushed forth into a solution that involved an apparatus consisting of a plush forehead cushion, attached to a removable mounting piece to be placed on the wall directly above the commode.

Flush with excitement and confident that he had discovered an urgent need in the marketplace, he received a patent for the device in 2002.

It has since expired.

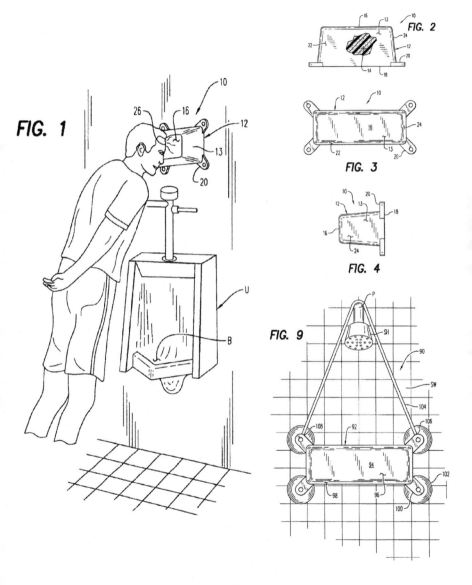

FIG. 1

FIG. 2

FIG. 3

FIG. 4

FIG. 9

## Female Urination Device

US20090089919A1
Inventor: *Cynthia K. Rudolph*
Patent Granted: *April 9, 2009*

Men have it easy. When the call of nature becomes their number-one priority, their options are nearly endless. If they are outdoors (i.e., enjoying a frosty libation, playing softball, playing baseball, playing Frisbee, playing touch football, playing soldier), where an actual restroom is inconveniently distant, all they need do is take a casual saunter around the perimeter of their immediate area, find a secluded spot (optional, depending upon the amount of libation consumed), then enjoy targeting and decimating the local ant population.

Women have it tough. When daily life, motherhood, and career combine to leave them stranded far away from civilization, they have few options available for relief. Sadly, an actual women's restroom offers little respite, often requiring the already stressed occupant to assume an uncomfortable, inelegant, squatting/hovering position so as not to make contact with the toilet seat. This time-consuming practice is particularly stressful at concerts and sporting events, when there are hundreds of women in line with exactly the same idea.

In this case, the old adage "Necessity is the mother of invention" holds particularly true. At least it did for Cynthia K. Rudolph, who saw an opportunity to even the score between the sexes when it came to this most basic of human endeavors. Her main aim was to re-choreograph the physical act by eliminating squatting altogether with the use of a clever, anatomically correct collection device terminating in a nozzle with which to direct the now organized stream in the direction of the user's choice.

For ladies, this meant they could now use any men's room with the same degree of confidence they felt when squatting behind a Dumpster in a parking lot. One small leak for women, one giant step for women's rights.

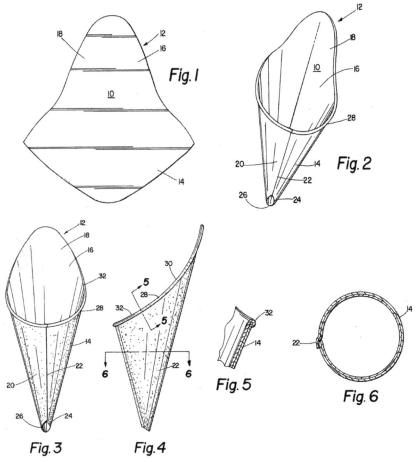

Fig. 1

Fig. 2

Fig. 3

Fig. 4

Fig. 5

Fig. 6

## Cardiac Pacemaker

US3057356A
Inventor: *Wilson Greatbatch*
Patent Granted: *October 9, 1962*

"He knew enough of the world to know that there is nothing in it better than the faithful service of the heart," said Charles Dickens.

As usual, Dickens had it right. Since antiquity, both the poets—and the scientists—have had one thing they can agree upon: The human heart is the fount of life, both spiritual and physical.

Aristotle saw the heart as "the source of all movement, since the heart links the soul with the organs of life." In 1797, Alexander von Humboldt found a dead bird in his garden and placed a blade of zinc in the beak and a shaft of silver into its rectum. An electric shock caused the bird to flap its wings and attempt to walk. He also tried the experiment on himself, which apparently induced him neither to fly nor to walk.

In the early 1950s, powered "portable" pacemakers were introduced as large bulky boxes filled with vacuum tubes that could not, of course, be implanted and were portable in name only, since they could only go as far as the nearest electrical outlet. At the least, this limitation ensured that the electric bill would be paid promptly each month.

In 1960, Wilson Greatbatch, an electrical engineer teaching at the University of Buffalo, was working on equipment to record unusual heart behaviors, when he accidentally discovered how to make an implantable pacemaker.

Greatbatch, a deeply religious man, described the event: "It was no accident, the Lord was working through me." He had taken the wrong part from his pocket, plugged it in, and was shocked to see that it worked! "I stared at the thing in disbelief," he said, immediately realizing that this small device could drive a human heart. He was right, it could and, eventually, it would.

But as author Zelda Fitzgerald once said, "Nobody has ever measured, not even poets, how much the heart can hold."

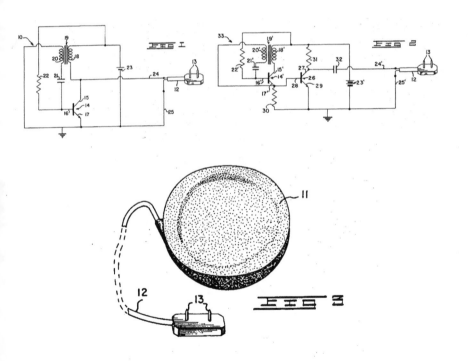

## Surgical Device to Prevent Nocturnal Emissions

US494437A

Inventor: *Frank Orth*

Patent Granted: *March 28, 1893*

We love rooting around the dusty annals of patented inventions to occasionally be rewarded with the discovery of a patent that breaks the mold, so to speak. So, when we exposed this device—which is actually put into operation by the onset of the wearer's tumescence—in a Rube Goldberg/Sigmund Freudian festival of sexual restraint, we couldn't resist.

In the inventor's own words: "Be it known that I, FRANK ORTH, of Astoria, in the county of Clatsop and State of Oregon, have invented a new and Improved Apparatus for Preventing Nocturnal Emissions, of which the following is a full, clear, and exact description.

"My invention relates to improvements in apparatus for preventing nocturnal emissions, and the object of my invention is to produce a simple apparatus which may be easily applied and worn, and which is automatically operated by an erection so as to cause the parts affected to be embraced by a chilling envelope which causes the erection to subside without a discharge."

We'll pass.

(No Model.)

F. ORTH.
SURGICAL APPLIANCE.

No. 494,437.                    Patented Mar. 28, 1893.

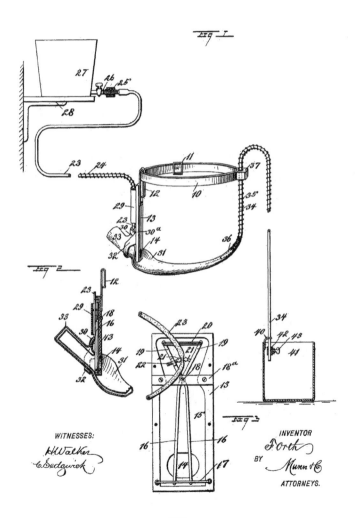

*Fig 1*

*Fig 2*

*Fig 3*

WITNESSES:
H.H. Walker
C. Sedgwick

INVENTOR
F. Orth
BY
Munn & Co.
ATTORNEYS.

## Life Expectancy Timepiece

US5031161A
Inventor: *David Kendrick*
Patent Granted: *July 9, 1991*

Everybody's got a bucket list based on their own idea of life expectancy. From learning how to make a perfect omelet to climbing Mount Everest, the lists are as varied as the people who dream them up. Movie hero Austin Powers wanted to "become an International Man of Mystery" and "earn Daddy's respect," but number one on most lists of lists is, you guessed it—to fall in love.

Of course, the determining factor in checking off items on any bucket list is time. There's never enough of it, but, more importantly, we don't know when our individual clock is going to run out. In 1858, according to *Scientific American*, life expectancy was based, to a large extent, on one's occupation. "Bank officers are the longest lived, their average age at death being sixty-eight to seventy-six, then clergymen and gentlemen, between fifty-five and sixty. Teachers are the shortest lived of all at thirty to thirty-five." Apparently, teaching school was considerably more hazardous to one's health in pre–Civil War America than it is today.

Then, lifestyle enters the equation. Actuarially, it stands to reason that a person who's never had a sick moment in their life, walks ten miles a day, and practices yoga in between meals of bean sprouts and tofu has a better shot at a long life than the guy stopping by for a triple cheeseburger on his way home for dinner.

Inventor David Kendrick wanted answers. He saw a need for a wearable chronographic measuring device that would display the hypothetical time remaining in the wearer's life. Clearly, this was much, much more easily stated than actually delivered. Imagine the sheer tonnage of variables that the Death Watch would have had to process to come up with anything like a reliable prediction—age, physical health, DNA, diet, socioeconomic factors—and those data points would have to be constantly reentered as time passed. The challenge was to finish your bucket list before you kicked the bucket.

The patent has expired (get it?).

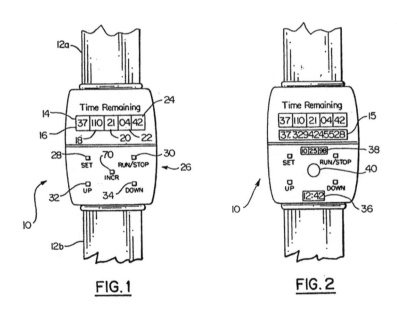

FIG. 1

FIG. 2

## Apparatus for Facilitating the Birth of a Child by Centrifugal Force

US3216423
Inventors: *George and Charlotte Blonsky*
Patent Granted: *November 1965*

The patent for this labor-saving device was filed in 1963 by George and Charlotte Blonsky, a New York City couple apparently intrigued by the possibility of post-natal flight.

Here's how the apparatus was designed to work. When a woman was ready to give birth, she would be strapped down on a circular, rotating steel table and spun at a speed approximating that of a 45 rpm record. Once sufficient rotational speed was reached, the projectile infant would be driven through the birth canal with force enough "to assist the under-equipped woman," as the patent abstract states, "by creating a gentle, evenly distributed, properly directed, precision-controlled force, that acts in unison with and supplements her own efforts." It isn't clear whether the woman was expected to be conscious by this point.

Few if any of the apparatuses appear to have been sold. As the website Improbable Research notes in its review, "Few people other than the Blonskys perceived the need for it. Their method stands rather outside most birthing traditions. Their mechanism is expensive and complex. Also, the tiny net designed to catch the child may be inadequate to the task."

Good last point. Say, for example, the apparatus facilitated the birth of a ten-pounder. The newborn could easily be delivered with enough force for it to shoot through the bottom of the net and land not in its mother's expectant arms but somewhere in New Jersey.

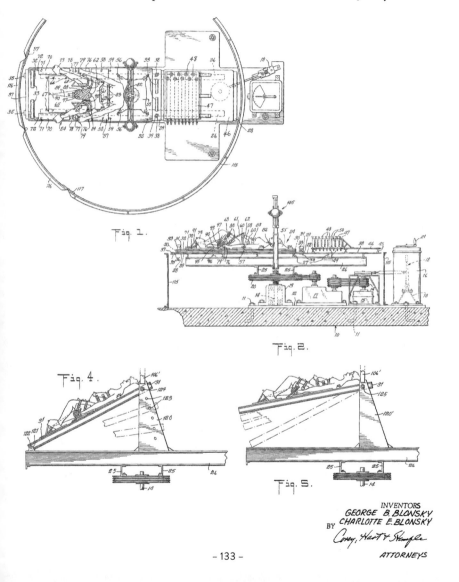

Fig. 1.

Fig. 2.

Fig. 4.

Fig. 5.

INVENTORS
*GEORGE B. BLONSKY*
BY *CHARLOTTE E. BLONSKY*

*Coney, Heart & Stample*

*ATTORNEYS*

—

## CHAPTER FIVE

# Fashion and Clothing

In eighteenth-century France, ladies at the court of Louis XIV wore so many layers of petticoats, overdresses, and massive hooped skirts that their outfits could weigh north of thirty-five pounds. Since it was deemed unhealthy and unfashionable to bathe, the courtiers implemented ivory scrapers with which to remove their sweat—a critical activity, we imagine, on a hot August day. In Edwardian times, manners of the day required gentlemen to tip their hats to gentle ladies so many times on a single block that someone felt the need to invent a Self-Tipping Hat to mechanize and automate the process. As playwright Tennessee Williams observed, "There comes a time when you look into the mirror and you realize that what you see is all that you will ever be. And then you accept it. Or you stop looking in mirrors."

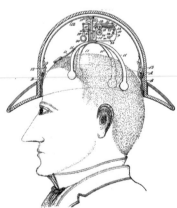

There's little risk of rudeness when one wears a "Self-Tipping Hat."

## Sled-Free Sledding Pants

US5573256A
Inventor: *Brent L. Farley*
Patent Granted: *November 12, 1996*

Inconvenience has been the motivating factor in countless inventions and patents.

How many times, for instance, have you found yourself at the bottom of a steep, snow-covered hill and wished you didn't have to carry your sled back up to the top? (Probably not many, actually, at least not in a long time, and in all likelihood never again.) You can't ask your mother to do it: You're thirty! But for Brent L. Farley, once, as a child, may have been one too many times. He never forgot the "inconvenience" of the situation, apparently, and so was inspired to patent this antidote to the childhood trauma.

As explained in the application for this device, "... it is particularly burdensome on the individual to drag or carry a sled or toboggan (hereinafter sled[s]) to the top of a hill after a long downhill run. The effort expended dragging or carrying the sled is known to be tiresome and detracts from the enjoyment of the sledding activity."

According to the application's abstract, Farley's invention works like this: "The sled comprises a seat component and a leg component" [good components to have if you're a pair of pants!], "both of which are ergonomically configured to correspond to the shape of the human body. The leg component is pivotally attached to the seat component to enable it to reciprocate from an upward walking position to a downward sledding position. The leg component includes an upturned distal [we had to look it up: away from the point of attachment] portion to allow the device to ride over small obstacles and bumps on the sledding path."

Realistically, the invention offers diminishing returns, given the warming of the planet. But, metaphorically, there may be something here for us: When life's obstacles and bumps leave us feeling as if we're at the bottom of the hill, anything that helps ease the way back up to the top may be worth wearing.

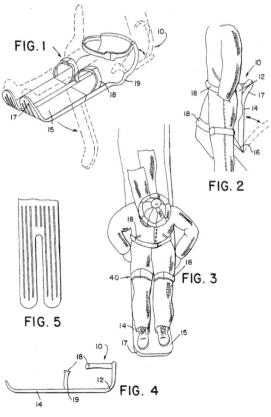

## Beerbrella

US6637447B2

Inventors: *Mason Schott McMullen (et al.)*

Patent Granted: *June 28, 2003*

It has been noted in the US Constitution that Congress may provide for the issuance of patents to further the progress of science and the useful arts. This is quite a broad manifesto, as almost anything that can be described as useful can be patented in the United States.

Mason Schott McMullen wasn't thinking about the Constitution that hot summer day when, mere minutes after popping open a can of frosty suds, the contents had warmed to a tepid, disappointing remembrance of a cold beer. Unacceptable, he thought. Attention must be paid to this dilemma.

Granted, the koozie had been around for a while, but it only served to insulate the contents of the can and was no match for the blazing sun on a July day. It soon became apparent that the solution lay not in keeping the cold in, but in keeping the sun out. And where does one find respite from the brutal rays of the sun? In the shade.

So, in the eons-long history of methods used to mitigate the power of old sol, from a shade tree, to a palm frond, to a parasol, to a Panama hat, the Beerbrella found its rightful place. Easily clipped onto a can, adjustable, and efficient, it worked. Bonus: Like almost every beach umbrella extant, it was a natural mini-billboard for advertising, promotion, and popular slogans like: "#1 Dad!" and "I'm with stupid!"

One problem arose. To drink from a Beerbrella'd can, it was necessary to move the little umbrella out of the way. Then, after sipping or, more likely, chugging, move it back into the sun defense position. This was fine for the first beer or two, but, we would guess, as the afternoon wore on, you'd find plenty of Beerbrellas over on that table next to the warm tuna-macaroni salad.

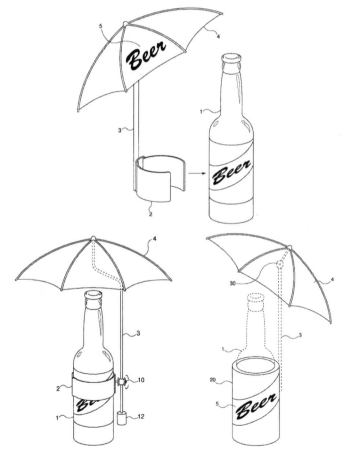

## Smoker's Hat

US4858627A
Inventor: *Walter C. Netschert*
Patent Granted: *August 22, 1989*

So addictive is nicotine that some people will do *anything* for a cigarette. They'll beg, borrow, steal, and spend their last dime. Stand outside in the rain and snow. Risk heart disease and lung cancer. They'll even wear something as ridiculous as this on their heads!

So it's not surprising that, in the late 1980s, as the nation began heeding warnings on the dangers of smoking, Walter Netschert turned the other way—to exhale! The California engineer–turned–inventor began smoking in college after a professor who questioned the reason for his mediocre marks suggested cigarettes to calm his student's nerves, which were negatively affecting his concentration. "It made all the difference in the world," Netschert later told a *Los Angeles Times* reporter, and saw no reason to stop now.

With the antismoking movement gaining momentum, Netschert devised a way to keep smoking while respecting the rights of non-smokers. His battery-powered Smoker's Hat consists of a clear plastic shield that forms an isolation zone in front of the face, complete with an ashtray and a clip that will hold either a cigarette or a cigar. A miniature fan pulls exhaled smoke through filters, removing smoke particles and odor, before returning purified air to the surrounding area.

The Smoker's Hat was a compromise of sorts, but it didn't extinguish its inventor's contrarian spirit. Nor did public ridicule: When Netschert was a guest on *The Tonight Show*, Johnny Carson reportedly told him, "You look ridiculous. You do know that, don't you?" Unfazed, Netschert dubbed himself Captain Huff-and-Puff and called those on the other side of his hat shield FAFs—Fresh Air Freaks.

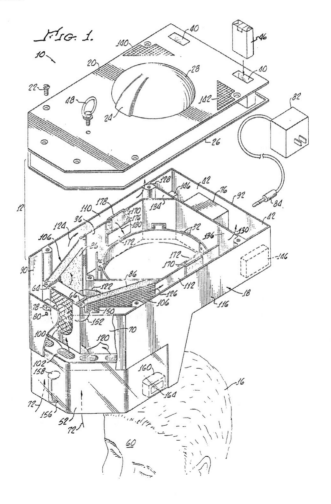

## Self-Tipping Hat

US556248
Inventor: *James C. Boyle*
Patent Granted: *March 10, 1896*

Manners: A term referring to the rigid rules of conduct governing correct behavior among members of a particular social group. In the super-starchy Victorian era, "good form" in society required assiduous adherence to a nearly endless list of social rules that today would garner comments like these: "You're kidding me, right?" and "What are you, nuts?"

Etiquette manuals of the time instructed gentlemen to lavish attention on the ladies present at all cost, putting aside their own needs to act as servants, guides, or even waiters.

The closest parallel for today's man would be the idea of his happily handing over control of the remote. But, in those days, all the rules were in full effect, and specified in a meticulous manifesto called *Frosts Laws and By-Laws of American Society* (1869). Here's how *Frosts* spells out the guidelines for a man to tip his hat in public:

"If a gentleman meets a gentleman, he may salute him by touching his hat without removing it, but if a lady be with either gentleman, both hats must be lifted in salutation. If a gentleman stops to speak to a lady, in the street, he must hold his hat in his hand during the interview, unless she requests him to replace it. With a gentleman friend etiquette does not require this formality."

In 1896, James C. Boyle had apparently become so exhausted by the physical strain and mental anguish involved in constantly tipping his hat to passing ladies on the boulevard that he was driven to create a mechanical device to perform this odious task on his behalf— mechanically. Thus, he designed a complicated clockwork mechanism, mounted inside a hat, to physically activate a tipping function.

There was, however, at least one practical drawback to his design. Activating the machine required bending forward in a bow to set a pendulum swinging, the weight of which would power the internal mechanism. So now, instead of simply saluting pedestrians with a quick tip o' the hat, Boyle needed to first stop and take a bow. Then, just stand there while his headgear proved him a gentleman above reproach.

J. C. BOYLE.
SALUTING DEVICE.

## Smart Ring

US20150277559A1
Inventors: *Marcos Regis Vescovi, Marcel van Os*
Assignee: *Microsoft Technology Licensing, LLC*
Patent Filed: *October 15, 2019*

Had we been even a little smarter, we would've guessed that once the closet started filling up with smart clothes—denim jackets wired to Bluetooth (Levi/Google), yoga pants that give feedback on poses (Nadi X), Smart Sneakers that tie themselves (Nike; see page 146)—smart jewelry couldn't be far behind. And we would've been right!

In October 2019, Apple filed an initial patent for a ring-worn device with touch-sensitive display and microphones for Siri-capability. Other smart rings, like the Oura Ring and the Amazon Echo Loop, are already on the market. But because Apple is always one for giving the competition a run for its money, since 2019 Apple's ring has already gotten smarter—and bossier.

The following year, the company refreshed the original patent application with a new version that expands the potential of the first, including gesture controls and a point detection function: In other words, the wearer could simply point to other Apple devices and connect.

But the Smart Ring (tech analysts suggest that's what the thing will be called) doesn't stop there in ordering other gadgets around. The latest application also suggests that the piece of jewelry could control home appliances, such as thermostats, light dimmers, and possibly audio volume levels.

At last check, in May 2020, Apple reported on Patentlyapple.com that the ring might expand to the length of the finger, enabling the device to contain more sensors and perform more functions.

Next up—at least we hope—the Wiseguy Pinky Ring: "Yeah, I'm pointin' at youse. Come ova here!"

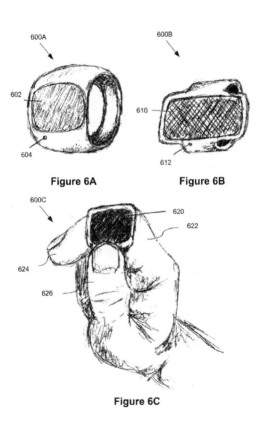

Figure 6A          Figure 6B

Figure 6C

## Smart Sneakers (AdaptBB motorized self-lacing basketball sneakers)

US20190090589A1

Inventors: *Thomas J. Rushbrook, Tiffany A. Beers*

Assignee: *Nike, Inc.*

Patent Granted: *March 28, 2019*

---

The history of the sneaker is the history of numerous nations, technologies, and industries. In fact, it's one of the most significant cross-cultural icons of the last century.

The British started the craze back in the 1830s when the Liverpool Rubber Company slapped rubber soles onto canvas uppers and called them "sand shoes." (No one would've called these things "smart"; there was no right foot or left foot, for example.) In the late-1890s on this continent, the US Rubber Company brought out sturdier kicks, called Keds, and near the end of World War I, when they were mass-produced, they picked up the nickname *sneakers*. As the story goes, they were so quiet that one could "sneak" around in them unnoticed.

That same year, Marquis Converse produced the Converse All-Stars. Endorsed by an Indiana basketball legend of the same name, Chuck Taylors became the best-selling basketball shoes of all time. And a year later, a German named Adi Dassler created a cooler version and named the brand and shoes after himself—Adidas—which Jesse Owens wore to win four gold medals at the 1936 Olympics.

Now Nike, the current King of Kicks, is catering to those either too busy or too lazy to tie their own shoes. The company's patent for the AdaptBB features a motorized device to tighten and loosen the laces. Used with the Nike Adapt App, wearers will be able to adjust the laces from a smartphone according to foot shape, size, and, says one product reviewer, "based on what you may be doing." Which suggests some interesting possibilities. Harking back to the original idea of "sneaking" around, if, say, you're doing something you maybe shouldn't be doing and need to make a fast but quiet exit, you won't want to have to stop to tighten pesky laces!

We're smart enough to get this. But for our money—the AdaptBB is priced at $350—we're waiting for someone to design a sneaker that will run for us!

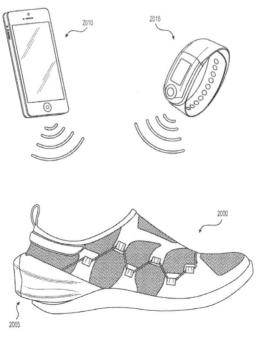

## Multiple Person Outfit

US5970518
Inventor: *Aurellius M. Jordan*
Patent Granted: *October 16, 1999*

This cozy "outfit/costume" features "three or more legs and the center leg will hold a leg of each person. It also has two arm sleeves, one for each person, inner sleeves that will hold the other two inner arms of the persons, and at least two neck apertures for the persons wearing the outfit/costume."

Note the design possibilities embedded in the patent description: "At least two neck apertures," which takes into account situations where another neck (and head and body, presumably) pops up at a party or for an intervention. Needless to say, perhaps, the garment isn't recommended when social distancing is required—unless, of course, one wishes to challenge the six-foot mandate. In such instances, it would be considered essential streetwear for social protests.

The benefits of the Multiple Person Outfit are obvious: significant savings on the cost of shopping for twins; reconciliation therapy for estranged couples in marriage counseling; and emergency warming for skaters who've fallen through the ice and may be in danger of hypothermia.

Regrettably, the garb's major shortcoming is also quite obvious: Although ample consideration was given conjoined twins, none was given those with Dissociative Identity Disorder—at least those with three or more personalities. Hopefully, an outfit designed for them is next up on the drawing board.

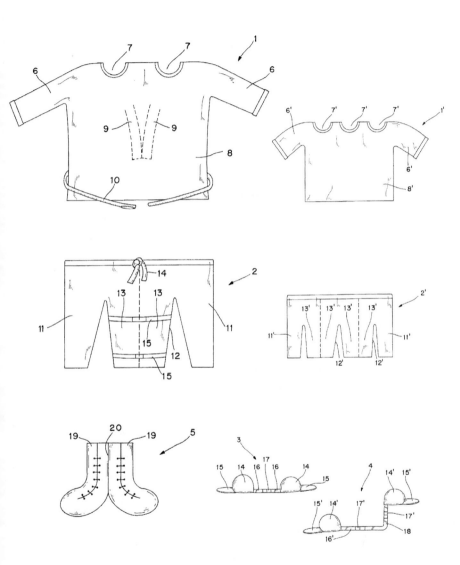

## Coat That Converts to Sleeping Bag
### (article of apparel convertible to a sleeping bag)

Inventor: *Veronika Scott*
Patent Pending: *2020*

Most of us don't have to worry about where we will sleep tonight. We'll kiss our loved ones goodnight, turn out the light, and climb into bed beneath clean sheets and warm blankets. But nearly 200,000 Americans sleep outside on any given night, and without shelter and sufficient bedding, they are prey to the elements, especially the cold.

Advocates for the homeless are working tirelessly on policies to address homelessness in America. It is inventors, though, who are patenting innovative products to help the country's silent population. Veronika Scott is one of them. As a college junior in Detroit in 2010, she received a class assignment to design a product that would serve the community in a material way. Having grown up in poverty, she immediately thought of the homeless and, while visiting a warming center that winter, conceived of a coat that would both warm and "empower" its wearers.

The EMPWR coat is a long storm coat with a hood and pouch in back that unfolds to enclose the feet and envelops and protects the wearer. Donations have fortified some coats. General Motors, for example, gave Scott enough scrap sound-deadening material for four hundred coats, adding insulation. Homeless men and women have been hired to sew the garments.

Other patents pertain to outerwear that converts to tents and temporary shelters (see facing page). Some include backpacks to carry additional bedding. While there appears to be no ready solution to America's growing homeless population, inventors like Veronika Scott are making their nights softer and warmer.

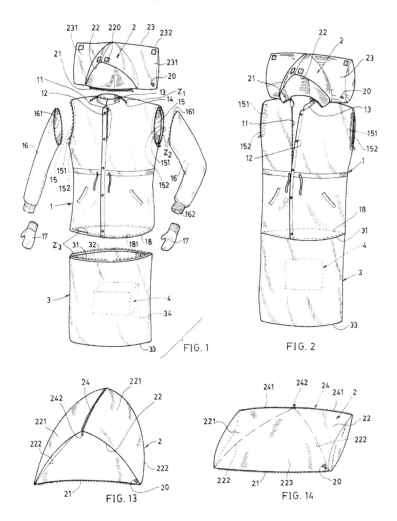

FIG. 1

FIG. 2

FIG. 13

FIG. 14

## Invisibility Cloak

US8253639
Inventor: *Nathan Cohen*
Assignee: *Fractal Antenna Systems, Inc.*
Patent Granted: *August 28, 2012*

In *Harry Potter and the Philosopher's Stone*, the first book in J. K. Rowling's epic fantasy heptalogy, young Harry inherits a Cloak of Invisibility to evade evildoers. The book came out in 1997, but the concept is hardly novel: Exactly a century before, H. G. Wells introduced the idea in his sci-fi classic *The Invisible Man*. In reality, in 2012, Fractal Antenna Systems, Inc., applied for and was granted a patent for just such a garment for muggles (read the series, okay?).

As with the descriptions of so many of the patents referenced in these pages, we're really not sure how this thing is supposed to work. In Claim 1 on the application, for example, the cloak is described, in part, as "An electrical resonator system, comprising a plurality of concentric electrical resonator shells, each shell including a substrate having first and second surfaces and a close-packed arrangement of electrically conductive material formed on the first surface, wherein the closed-packed arrangement comprises a plurality of self-similar electrical resonator shapes and is configured to operate at a desired passband of electromagnetic radiation . . ." See what we mean?

We're also in the dark as to the product's availability in the marketplace. What we have seen, however, are facsimiles of the black cloak with green lining that Harry wears: They're manufactured by Wow! Stuff, the company that markets Lumos Wands, Golden Snitch

Heli Balls, and other items found in Rowling's mega-series. The Standard Invisible Cloak, which retails for between $39.99 and $70.00, comes with a downloadable app and authentication code that allow wearers to reenact their favorite scenes from the films based on the books.

Note: Don this item in the store if you wish, but if you walk out without paying for it, you'll be totally visible!

Now you see him, soon you won't.

| Bra (backless brassiere) |
|---|
| US1115674 |
| Inventor: *Mary Phelps Jacob* |
| Patent Granted: *November 3, 1914* |

Women have always been slaves to fashion. But just before World War I, Mary Phelps Jacob freed them from 350 years of enslavement to one particularly painful fashionable convention—the dreaded corset.

Jacob was more independent that most women of her day. Wealthy and privileged, she was first married to a descendent of one of Boston's most prominent families, then to the scion of a wealthy and even more socially prominent Boston family.

One evening before dressing in a sheer gown bought for a social event, she got into a corset—a long, rigid undergarment that encircled a woman's ribs and used whalebone stays and laces to compress the waist until her figure resembled an hourglass. (Introduced to womankind by Catherine de Médicis of France in the 1550s, the corset was once referred to by *British Vogue* as "a patriarchal instrument of torture that deformed the female body.")

When the corset Jacob put on showed above the gown's plunging neckline, she had had enough. According to the website Phelpsfamilyhistory.com, she threw away the contraption, and, with the help of her maid, sewed two silk handkerchiefs together, then tied them behind her back with lengths of pink ribbon and cord. The ribbon was later replaced by thin nylon straps.

Jacob was granted a patent for the design in 1914, but preferring nightlife to running a business, she sold the patent the next year to the

Warner Brothers Corset Company (now Warnaco) for $1,500 (today's equivalent of $25,000). Bra sales slipped in the late 1960s as young women discarded their bras—some burned them—as a symbol of the women's liberation movement.

Today, bras are back on and part of a $7.2 billion market.

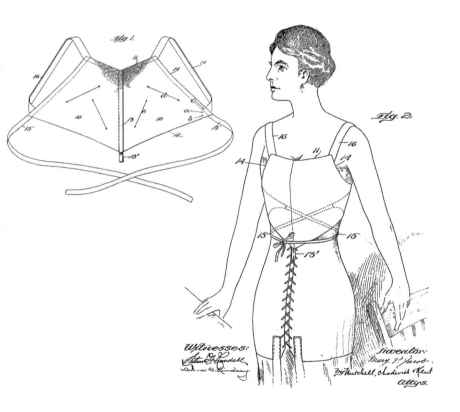

## Inner Pocketed Two-Person Jacket

US4773102A
Inventor: *Darryl A. Curtis*
Patent Granted: *September 27, 1988*

To be warm, dry, and close to your social circle—these are three of the conditions required for sustaining human life, especially in northern climes. Since the icy days of the wooly mammoth, whose coat was prized as the ultimate in winter wear if you and your tribe could bring one down with a slingshot and a rock, men and women have sought relief from winter's chill. When the hominids first walked out of the relentless crucible of African heat, they soon discovered this freezing new threat to their already tenuous existence. And they learned to adapt.

Along the way, over the millennia, man and his chosen partner have bundled up together in the fur of any animal that had fur. From sheep to seals, from beavers to bison, from rabbits to coyotes, if it had fur, it was used to make a coat or a blanket.

Then, as always, science joined the party with the inception of synthetic fibers, whose miraculous properties boasted warmth without weight and economical manufacture.

Inventor Darryl A. Curtis must have seen the technical advances in cold-weather apparel as necessary progress. Parkas, snowsuits, and arctic survival gear were all well and good, but what about the human connection that comes with close proximity to a chosen partner? Why not create a modern winter jacket tailored for two people to wear at the same time?

Let's examine a hypothetical scenario:

It's a January morning in Minnesota. We're in the mudroom of George and Mary Antrobus.

Mary: "Did you walk the dog?'

John: "It's your turn."

Mary: "No, it's not. I did the dishes, you walk the dog."

John: "See how you are?!"

(They pull their two-person jacket off its hook and solemnly, angrily zip themselves into it.)

Mary: "Let's go."

John: "Don't tell me what to do!"

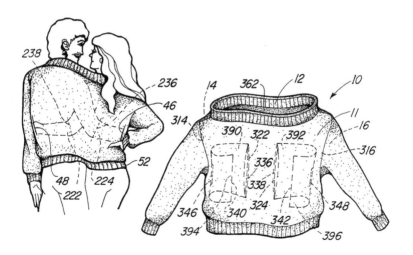

## CHAPTER SIX

—

# Games and Toys

How could we have grown up without many of these stars of the game and toy universe? The simple answer is that we never did, which is why we love the patents in this section. Beat us with a stick (or with the Dog Toy on page 170), but, to us, LEGOs are every bit as ingenious as the Microwave Oven or the Windshield Wiper System with Intermittent Operation (okay, the last one's close). It's also the stories behind these patents that gripped our imaginations. The accidental tumble off the shelf of a coil of wire that slinked down the stairs? The NASA experiment for a spacecraft that turned into a high-powered water gun? The iconic collectible of American girlhood, inspired by a German escort doll? Come on! You can't make this stuff up—and we didn't have to, thanks to the comedic genius of their inventors. Thank you, one and all!

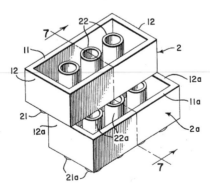

Why do humans enjoy snapping one piece of plastic to another? Who knows and who cares? One hundred million LEGO bricks are sold each year!

## LEGOs (toy building bricks)

US3005282A
Inventor: *Godtfred Kirk Christiansen*
Assignee: *The Lego Group*
Patent Granted: *October 24, 1961*

"If necessity is the Mother of invention, then play is its Father," argues Steven Johnson, author of *Where Good Ideas Come From* and *Wonderland: How Play Made the Modern World.* A sense of playfulness, often of mindless amusement, is a trait shared by many of the most successful inventors and innovators in history. In the case of LEGOs, the iconic toy manufacturer, fire—an archetypal symbol of inspiration and regeneration—has played almost as important a role.

Godtfred Kirk Christiansen, a modestly successful but ambitious carpenter, was making ladders, stools, and ironing boards in his shop in Billund, an isolated village in Norway, in 1916. His business grew out of his love of whittling and playing with wood. One day, his twelve-year-old son accidentally lit a pile of kindling on fire and the shop, along with the family home, went up in smoke.

Undaunted, Christiansen built a larger shop and began making wooden toys. When that shop, too, burned to the ground, he built an even larger shop to make the wood ducks whose bills opened when pulled on a string. And when a *third* fire leveled the plant and World War II led to the rationing of wood, he bought Denmark's first plastic-injection molding machine. By 1949, his company was turning out interlocking plastic blocks called the Automatic Binding Brick.

Today, the Lego Group sells close to 100 billion LEGO bricks a year. They are used in hundreds of lines, from architecture to race cars, as well as kits based on books and films, such as *Harry Potter* and *Lord of the Rings*, and once-remote Billund has become a popular tourist destination. The company estimates that since its founding, enough bricks have been manufactured to build 46,826 average-sized homes.

In the near future, perhaps your house will be built entirely of LEGOs!

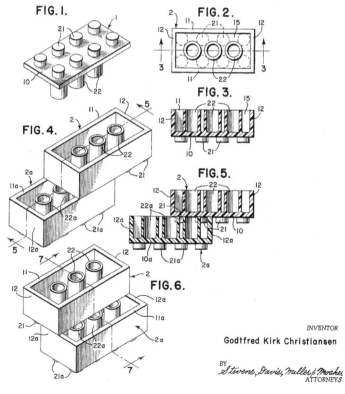

INVENTOR

**Godtfred Kirk Christiansen**

BY
*Stevens, Davis, Miller & Mosher*
ATTORNEYS

## Barbie

US3009284A
Inventor: *John W. Ryan*
Patent Granted: *July 7, 1959*

She is the quintessential American woman—with a colorful past. Her full name is Barbara Millicent Roberts, but the world knows her as Barbie.

She was "born" March 9, 1959, the day Mattel president and Barbie inventor Ruth Handler debuted the doll at the American International Toy Fair in New York. She had been conceived when Ruth noticed that her daughter, Barbara, and her friends preferred playing with dolls that looked like adults, rather than babies. Ruth's husband, Elliot Handler, took the idea of such a doll to his all-male ad executive team at Mattel, who rejected it because they saw little mass appeal for such a doll. Well done as usual, gentlemen. (Perhaps it was their wives who convinced them otherwise!)

Ruth had her own idea for the doll's look, based on a risqué German doll, *Bild Lily*, created in 1952. A high-end call girl, she was featured in the comic strip *Bild-Zeitung*. She was originally marketed in the United States as a Teenage Fashion Model and 350,000 dolls were sold the first year.

Let's not forget the boyfriend, Ken, who showed up in 1961 and hung around until 2004, when they broke it off. That's when a new guy named Blaine Gordon, a surfboarder from Australia, threw his shrimp on the Barbie. He lasted until Valentine's Day 2011, when Barbie told friends that her "heart only beats for Ken."

Successful? You have no idea. Since Mattel created her, succeeding generations have rapturously swooned over her feminine charms. An ingenue in perpetuity, by her fiftieth birthday in 2009, over one billion Barbies had been sold by Mattel. Every three seconds, a Barbie doll is sold somewhere in the world.

When it comes to financial independence, Barbie's all set. A mint condition Barbie #1 sold at auction in 2006 for $27,450. It has been estimated that Barbie's Dream House, with its seven rooms, an elevator, a garage, and a pool, would be worth around $16 million. If it were real.

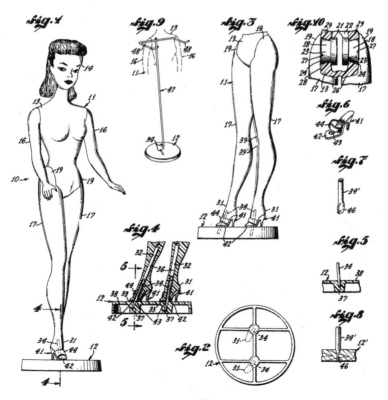

## Angel Action Figure

US5588895
Inventor: *Diana A. Larsen*
Patent Granted: *December 31, 1996*

Divine intervention may have played a part in this invention. Or maybe it was simply the inventor putting her faith into action. In any event, something inspired Diana Larsen to patent her Angel Action Figure because here it is—kneeling, praying, and ready to kick Satan's butt.

That's because, like all action figures, the angel with an "H" (for *Heaven?*) emblazoned on his chest has numerous movable parts, including a set of wings. The doll was designed, according to the application copy, "to simulate a conventional representation of an immortal, spiritual being superior to humans, that is an attendant and messenger of God, which can be used as a plaything of a child."

But play doesn't appear to have been foremost in Larsen's mind at the time. One intention of the Mount Ayr, Iowa, resident's invention was to "overcome the shortcomings of the prior art devices." That would cast her angelic doll against the likes of G.I. Joe, Hulk Hogan, Rowdy Roddy Piper, and, especially, that heathen action figure Hellboy!

In addition to a set of upper *and* lower wings, the Angel Action Figure comes with a basketball-like team jersey, booties, star ankle joints, and its own halo, obviating the need to go into a toy store and ask if they have any. And thank God for that!

## G.I. Joe Action Figure (toy figure having movable joints)

US3277602
Inventors: *Samuel F. Speers, Hubert P. O'Connor*
Assignee: *Hasbro*
Patent Granted: *October 11, 1966*

If Barbie had been in the market for a boyfriend back in the sixties, she might have done worse than G.I. Joe, a strapping figure of . . . a figure. When called to duty in 1964, the eleven-and-a-half-inch-tall hunk of plastic was the most popular action figure in America (he also happened to be the first)! What girl wouldn't want to date him? What boy wouldn't want him in his toy chest? By the end of the decade, plenty, as it turns out.

Writing in *Smithsonian* magazine, author Jimmy Stamp traced the figure's lineage to the Hassenfeld brothers, former pencil makers whose company, called Hasbro, became one of the largest toy makers in the world. Don Levine, the company's director of marketing and development at the time, took note of Mattel's success with Barbie (see page 162) and figured boys needed a macho counterpart.

But Levine was also inspired by an artist's scaled wooden mannequin that had movable parts. That inspired the nineteen "points of articulation"—movable joints and body parts—that replicated human anatomy and proved key to G.I. Joe's success. (The name came from the 1945 film *The Story of G.I. Joe*, about war correspondent Ernie Pyle.)

Hasbro had scored a hit with Mr. Potato Head (see page 172) in 1952, but that was eclipsed by G.I. Joe. Within two years, sales of the toy accounted for two-thirds of the company's profits. Other action figures quickly followed, representing the other branches of the military—Joe was a Marine—each with fatigues, combat boots, a cap, and dog tags.

G.I. Joe was retired from active duty in 1968, victim not of the Vietnam War but of antiwar protests and sentiments. And had they been a couple, Barbie would've surely dumped him by the mid-'70s for the Six Million Dollar Man action figure!

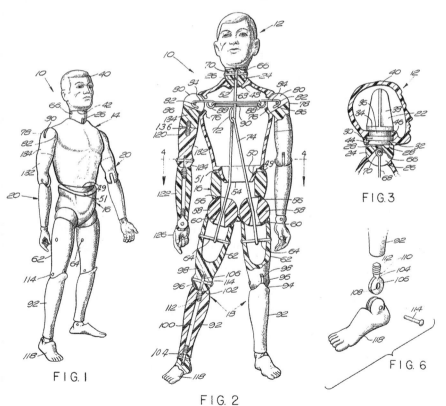

FIG. 1

FIG. 2

FIG. 3

FIG. 6

## PEZ Dispenser (pocket article dispensing container)

US2620061

Inventors: *Oscar Uxa, Eduard Haas*

Patent Granted: *December 2, 1949*

First came the candy: tiny, brick-shaped mints, invented by Austrian Eduard Haas in an attempt to help his countrymen and -women quit smoking (or at least mask the smell of smoke on their breath). He named it PEZ, short for the German word for "peppermint" —*pfefferminz*—and labeled the wrappers in what is known as the "PEZ font": letters formed from the shape of the candies!

Americans weren't necessarily interested in quitting smoking or in sucking on mints. But they did like the shape of Haas's candies, as well as the idea of sweet, flavored lozenges. Most of all, they went crazy for the PEZ dispenser that fellow Austrian Oscar Uxa introduced to the States in 1952.

Designed to resemble a cigarette lighter, the narrow case held twelve candies, stacked on top of a spring-loaded cartridge and released with the flick of a lever at the top. Three years later, in a shameless marketing ploy to children and silly adults, the company stuck heads on the dispensers, starting with Santa Claus and Mickey Mouse.

Today, the company Haas built sells nearly four billion PEZ candies in ten flavors a year, and has created over fifteen hundred different novelty dispensers, with collectors paying tens of thousands of dollars for rare offerings.

In 1973, the company built a manufacturing facility in Orange, Connecticut, where twelve million of the hard confectionary bricks are produced every day. The facility also houses the PEZ Visitor Center and the largest display of PEZ dispensers in the world.

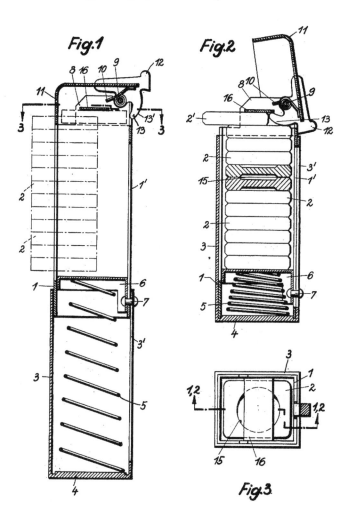

**Dog Toy**

US636069391
Inventor: *Ross Eugene Long III*
Patent Granted: *March 26, 2002*

The genius of this patent lies in its utter simplicity. While not exactly pioneering, it is also one of those ideas for a new product that makes us slap our heads and exclaim, "Why didn't we think of this?!"

The abstract for the patent says—actually we're not sure what it says, but here's how it reads in part: "An animal toy, comprising: A) a solid main section having a diameter and a longitudinal length and extending a predetermined distance along said longitudinal length; and B) at least one protrusion attached at one end thereof said main section and extending a predetermined distance therefrom and wherein said at least one protrusion includes a second longitudinal axis that is not in parallel . . ."

In essence, as well as in every other possible sense, the Dog Toy is a stick. That's right, a stick. But not just any stick. This is a fake stick, designed to trick dogs and save owners/throwers the time and effort of prying a real stick from the canine's jaws. The "genius" behind the product is that it recognizes and solves a real problem: Dogs love chasing sticks, but they also love chewing them to smithereens, forcing us to find new sticks, again and again, until the landscape is stickless and our best friend stares at us with bewilderment and heartbreaking disappointment.

Should a dog sniff out the inauthenticity of the product? Not a problem. The renewable chewable can be given to a kid as a Christmas or birthday present with the instructions, "Now go outside and play, and don't come back in until dinnertime!" No worries about claims of patent infringement on this puppy: A patent for the groundbreaking product was granted in March 2002, but failure to pay the initial maintenance fee sent it into the public domain, where it remains to this day.

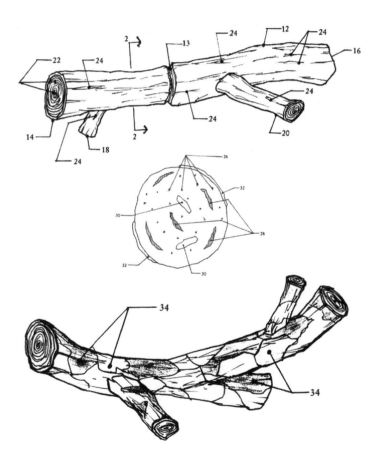

## Mr. Potato Head

US3660926
Inventor: *George Lerner*
Assignee: *Hasbro*
Patent Granted: *May 9, 1972*

It's basic. The potato is a root vegetable, native to the Americas, a starchy tuber of the plant *Solanum tuberosum*, a perennial of the nightshade family, *Solanaceae*.

But that's not what inventor George Lerner was thinking in 1949. He saw the humble spud as much more than a supporting player on a dinner plate. Rather, he saw it as a little three-dimensional stage set, upon which to portray the myriad facial varieties and expressions of humankind.

Or not. Maybe he thought it would just be fun to stick some plastic bits on a tater and have a giggle at the funny faces. Either way, Lerner hit one out of the park with Mr. Potato Head, a beloved toy that almost every kid has played with and every parent has cursed after stepping barefoot on one of its parts.

It's had big moments over the years, including a star turn on April 30, 1952, in the first television advertisement ever aimed at children. That commercial was produced by the manufacturers, the Hassenfeld brothers. The Hassenfelds earned more than $1 million in the first year of Mr. Potato Head production; in 1968, they changed their company name to Hasbro, and today they are the third largest toy company in the world. But don't worry about Lerner. He had a 5 percent royalty on every sale.

Over the years, it became obvious that Mr. Potato Head was lacking a family. Enter Mrs. Potato Head, their offspring—Yam and Spud—and the kids' friends: Kate the Carrot, Pete the Pepper, Oscar the Orange, and Cookie Cucumber.

We imagine family dinner at the Potato Head household, although loaded with carbs, might be a bit light in the meat department.

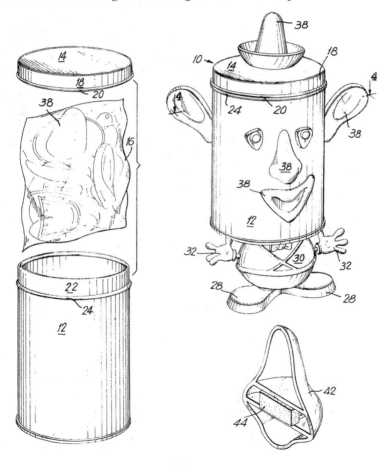

## Slinky (toy and process of use)

US2415012A

Inventor: *Richard T. James*

Patent Granted: *January 28, 1947*

For all the genius and obsessive effort behind many of the pioneering patents of the last four centuries, homage must be paid to that magic ingredient behind so many other amazing inventions and products: accident.

Richard James, a mechanical engineer in Philadelphia in the 1940s, was experimenting with wire coils to keep sensitive nautical equipment safe aboard ships at sea. Fortunately, he couldn't keep his own coils safe! As noted previously, when one of the coils accidentally fell off a shelf and practically walked itself down an incline, James and his wife Betty knew it was trying to tell them something: Your life is about to change!

According to the National Toy Hall of Fame—the Slinky was inducted in 2000—James invented a machine that could reduce an eighty-foot coil into a two-inch spiral. The couple then borrowed $500 to manufacture the first Slinkys.

The product's purpose, as described in the 1946 patent application: "To provide a helical spring toy which will transfer its turns from one end to the other in an entertaining manner."

And entertain it has. To date, over 350 million Slinkys have been sold, changing Richard and Betty's lives, though in more ways than they wished for. In 1960, Richard left the business—and his family—to join a religious cult in Bolivia. It was Betty who had come up with the name *Slinky*, and now it was she who brought the company out of near insolvency while raising the couple's six children.

## Rubik's Cube

US4378116
Inventor: *Ernő Rubik*
Patent Granted: *March 29, 1983*

You've likely held one and enjoyed the tactile clicks as you spun its multicolored faces to scramble the order of the individual tiles. But the fun (torture) begins when you attempt to return the tiles to their starting positions. It may take a while. There are forty-three quintillion possible configurations of the cube. It's estimated that only one person in twenty who owns one will ever solve it.

Inventor Ernő Rubik, who worked at the Department of Interior Design at the Academy of Applied Arts in Budapest in the mid-1970s, wasn't out to create the world's most popular toy when he constructed an early iteration of the cube as an exercise for his students. The challenge was to solve the three-dimensional structural problem of moving the parts independently without the entire mechanism self-destructing. He didn't even realize he had created a puzzle until the first time he scrambled his new cube and then tried to restore it.

The first test batches of the then-named Magic Cube were produced in late 1977 and debuted in toy shops to modest success. Soon after an appearance at Germany's Nuremberg Toy Fair, with its star on the rise, the cube was licensed by Ideal Toys for release worldwide. Ideal wanted a recognizable name to trademark, so the humble little puzzle was renamed after its inventor in 1980.

It's done pretty well over the years. More than 350 million have been sold worldwide. And Ernő Rubik, who was making $200 a month as a teacher, saw a slight increase in pay when the royalties kicked in. Let's call it roughly $30,000 a month.

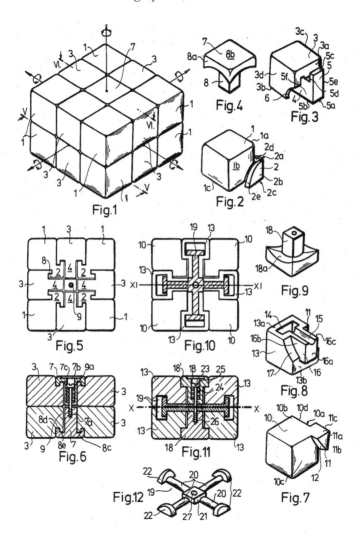

Fig.1

Fig.2

Fig.3

Fig.4

Fig.5

Fig.6

Fig.7

Fig.8

Fig.9

Fig.10

Fig.11

Fig.12

## Super Soaker (squirt gun)

US4591071
Inventor: *Lonnie G. Johnson*
Patent Granted: *May 27, 1986*

It doesn't take a rocket scientist to come up with an ingenious idea, but, in the case of this blockbuster toy, it helps.

Lonnie Johnson, one of the most prolific black inventors in history, with over 120 patents to his name, conceived the Super Soaker while working at NASA's Jet Propulsion Lab. He was working on the *Galileo* spacecraft, for which he invented a power supply mechanism, and also a heat pump that used water instead of Freon, a pollutant.

"I was experimenting with nozzles I'd made," Johnson told *Forbes*, "that shot a stream of water across the bathroom and I thought they'd make a good water gun." The trial marketing campaign for the toy? Lonnie's young daughter, to whom he gave the prototype, and her friends in the neighborhood, who gave it a thumbs-up.

But "water gun"? Water rocket-launcher is more like it! Compressed air in the chamber is further compressed by a pressurization pump. Then, as described in the application's abstract, "a battery-powered oscillator circuit and a water flow powered sound generator produce futuristic space ray gun sound effects when the gun is shooting."

Patented in 1986, Johnson licensed the device to Hasbro. To date, the Super Soaker has realized more than $1 billion in sales and is regularly listed as one of the top-selling toys in the world, theoretically threatening to soak the entire planet.

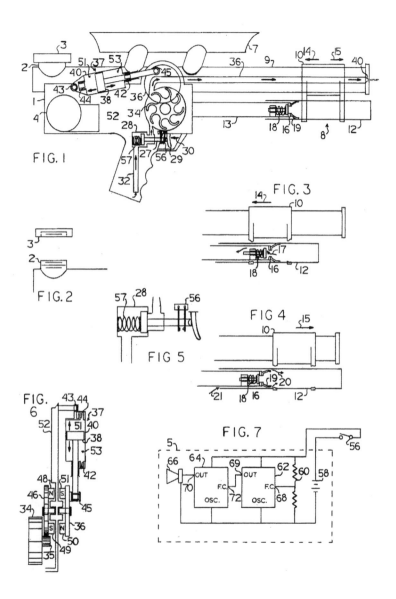

FIG. 1

FIG. 2

FIG. 3

FIG 4

FIG 5

FIG. 6

FIG. 7

## Natural Creeping Baby Doll

US18435
Inventor: *George P. Clarke (et al.)*
Patent Granted: *August 29, 1871*

At the top of the list—if there is one—of the cutest things in life, a baby making her first earnest attempts to crawl usually out-cutes even a Labrador puppy. Guaranteed to transfix even the most cynical of adults, this impossibly adorable infantile rite of passage has charmed kings and commoners, captains of industry, and babysitters into submission for thousands of years.

Nineteenth-century inventor George P. Clarke somehow must have seen an opportunity to mimic nature's greatest design by way of a whirring collection of gears, nuts, and bolts. Why? Good question. One reason might be that, at the time, clockwork automatons were in vogue and seen as things of wonder by society folks who could afford to stand around being amused.

Automatons have existed for centuries. One of the earliest examples on record is a life-size humanoid figure, complete with artificial organs and capable of singing and dancing, built in China in the tenth century BCE.

The drawing you see here presents the invention of a mechanical, articulated doll. It accompanied the patent submission for a Natural Creeping Baby Doll, a worrisome title at best. The doll's head, arms, and legs are made of painted plaster and hinged to a brass clockwork body that mimics a mechanical interpretation of crawling. But the doll actually moves forward by rolling along on two wheels. You can't get much cuter than that.

Overheard in a Victorian age front parlor: Husband: "What's that infernal clanking noise, dear?" Wife: "Oh, silly me, I forgot to change the baby's oil."

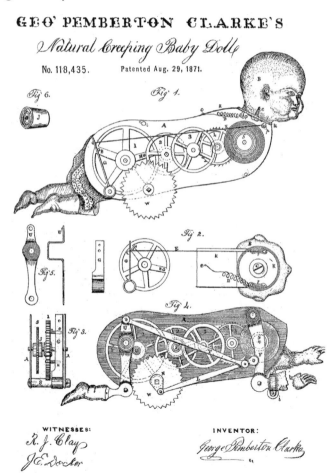

## Ouija Board

US446054
Inventors: *Elijah J. Bond, Kennard and William H. A. Maupin*
Patent Granted: *November 10, 1891*

Toy? Game? Hardly either one, especially if you're playing with the dead.

These days, Ouija boards are usually considered novelty items used to conduct séances and communicate with the afterworld. But when these "talking boards," as they were originally called, made their appearance in the United States in the mid- to late 1899s, people either listened raptly or ran screaming from the room.

So popular did Ouija boards become that a Museum of Talking Boards sprang up in Baltimore, Maryland. According to the museum's website, in 1848 two sisters in an upstate New York cabin said they'd contacted the spirit of a dead guy, thus launching a national craze. Soon after, mediums (intermediaries between the living and the dead) began placing their hands on tabletops, which would tilt in the direction of the sitter with whom a spirit wanted to make contact.

By 1890, Elijah Bond and Kennard and William H. A. Maupin—grifters of the kind Mark Twain caricatured in *The Adventures of Huckleberry Finn*—had gotten wind of the board's popularity and saw dollar signs. They patented them as Ouija boards. (The name is believed to be a combination of the word *yes* in French and something in German: *ji*?)

Made of wood, the boards feature two curving rows of the alphabet above a row of numbers, and the words *yes* and *no* in the upper-left and upper-right corners and *goodbye* at the bottom. In the center is a planchette, a heart-shaped device with wheels and a pencil, on which users place their hands and are "guided" to spell out names or messages.

Hundreds of Ouija and other talking boards are marketed to children and teens today. If they were our kids, we'd just want to know which deceased relative or friend they were "talking" to.

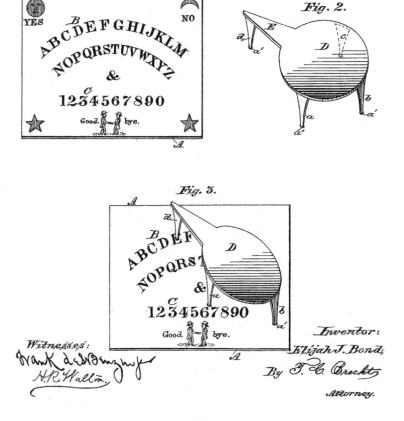

## CHAPTER SEVEN

—

# Home and Entertainment

There was a time, not so long ago, when, at the end of a long day's physical labor a family would gather around the hearth after the evening meal to read stories, tell tall tales, and gossip about the neighbors. Then the radio replaced the fireplace, substituting the snapping of fingers and the tapping of toes to hit songs for the crackle of firewood. Soon, television bullied its way onto the American scene and, for the first time, an eager nation got its news and entertainment on the same day it happened. But even that wasn't fast enough for Bill Gates and Steve Jobs. With the advent of the personal computer, information morphed into a commodity that could be bought and sold. And that comfy fireside chat with the family? It's now delivered to your wristwatch!

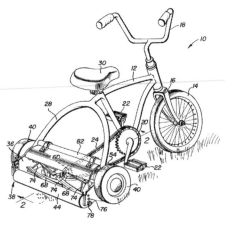

Happy sixth birthday! Now go outside and ride your new tricycle back and forth across the lawn. *Evenly!*

## Flushing Toilet

UK814

Inventor: *Alexander Cumming*

Patent Granted: *1775*

The year was 1775 in Edinburgh, Scotland. A ragtag bunch of unwashed insurrectionists in the American colonies had grown into a formidable fighting force—and a real pain in the ass for King George III.

Concurrently, in the British Isles, a Scottish watchmaker and inventor named Alexander Cumming was bearing down on a somewhat more intimate but equally pressing problem involving the posterior—how to mitigate the stench of sewer gases released during daily evacuations.

His efforts were rewarded when he (1) conceived of the S-Trap to retain water permanently within the waste pipe, and (2) improved the flush mechanism so that it actually flushed. However, in 1775 Scotland, one wonders where all that human waste was flushed to.

Today's modern, convenient, computerized, electronically controlled and heated appliances have become more like expensive entertainment systems than toilets until they are overfilled and the water threatens to overflow. In that panicky moment, rich or poor, we are all equal.

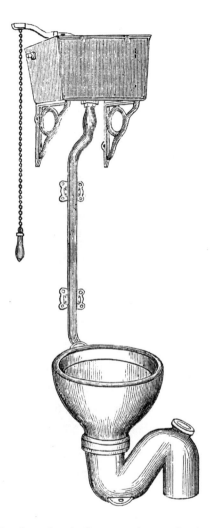

An early iteration of one of mankind's most popular scientific achievements.

## Motorized Rotating Fork

US6442846B1
Inventor: *Michael Maiello*
Patent Granted: *September 3, 2002*

Although the conventional narrative claims that Marco Polo introduced pasta to Italy, following his exploration of Asia in the late thirteenth century, pasta can be traced back as far as the fourth century BCE to an Etruscan tomb depicting a group of natives making what appears to be pasta. But as usual, the Chinese were ahead of everybody, making a noodle-like food as early as 3000 BCE. It would take nearly five thousand years, the invention of the telephone, and the adoption of an actual street address to make home delivery an option.

The term *spaghetti* (Italian for "long thin strings") first showed up in Sicily around the twelfth century CE, and its popularity spread throughout Italy after the establishment of spaghetti factories in the nineteenth century and, much later, in American Spaghetti Westerns, which disappointingly had nothing to do with pasta at all.

But for all its international popularity, the consumption of spaghetti, by dint of its shape, was hampered by a logistical problem: how to get it from the plate to the mouth without spraying everything or anyone nearby with a mist of the sauce ("gravy" in Tony Soprano's house) of your choice.

Michael Maiello came up with a solution. Instead of simply winding the strands against the inside of a spoon with a fork, a practice that had made gaining weight nearly effortless for hundreds of years, why not add an electric winding mechanism to the equation? Then, all eaters had to worry about was controlling the rpm of the fork, washing it in soap and water, and recharging the battery.

Anyone for tiramisu for dessert?

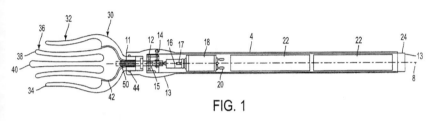

FIG. 1

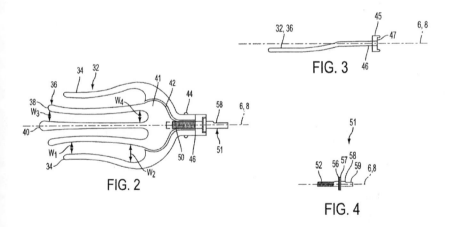

FIG. 3

FIG. 2

FIG. 4

## Light Bulb Changer

US6826983
Inventor: *Thomas Magdi*
Patent Granted: *December 7, 2004*

How many software engineers does it take to change a light bulb? None. That's a hardware issue. That thought must have been the impetus behind inventor Thomas Magdi's bright idea.

Thanks to good old Tom Edison and his crew at Menlo Park, we are surrounded, 24/7/365—above, behind, and below—by light bulbs. Since 1880, when commercial lighting was adopted on a large scale, the effect of nightfall and inky darkness on commerce, entertainment, navigation, and safety has been mitigated or eliminated to a large extent by the humble glass vacuum bulb with its warmly glowing filament.

But our man, Magdi, apparently had a big problem with little light bulbs. They burned out and they had to be replaced. This involved removing the burned-out bulb and screwing in another one. Elementary, right? Not to Magdi.

He set out on a personal quest to invent a way to change a light bulb without human intervention. Eventually, he devised an exquisitely complicated Rube Goldbergian flight of mechanical fantasy with as many parts as a fine watch. This device detected the moment a bulb burned out, then physically extricated the dead bulb and screwed in a new one.

There's one basic flaw here. When the contraption removes the bad bulb, what does it do with it? If it remains inside the machine, wouldn't one need to replace it with a new substitute bulb in advance of the next burnout? If that is the case, changing a light bulb has now attained a degree of difficulty similar to that of replacing the carburetor in your car every time you fill up your gas tank.

"Oh, honey, the front door light has burned out."

"I'll call the mechanic."

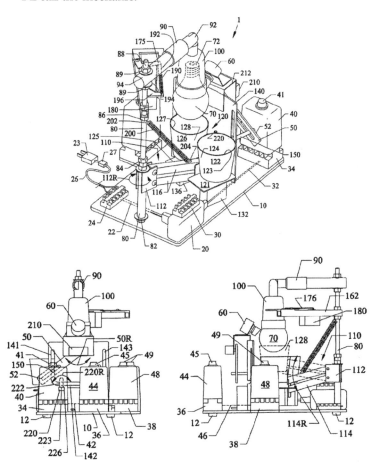

## Television (cathode ray tube)

US1773980
Inventor: *Philo Taylor Farnsworth II*
Patent Granted: *August 26, 1930*

It has delivered to us the best and worst news, and entertained us for most of a century. We remember exactly where we were when Neil Armstrong set foot on the moon, when JFK was assassinated, and when the Chicago Cubs finally won the World Series. When the announcer queried, "What time is it, boys and girls?" we knew it was "Howdy Doody Time!" and the medium that brought it into living rooms everywhere transformed the world in ways yet to be fully grasped.

Television is yet one more example of a technology and product built over decades of experimentation by numerous inventors and tinkerers. In the 1920s, isolated electronic whizzes in Russia, England, and the United States transmitted crude images. But the first clear images sent electronically through a TV screen were demonstrated here in 1927 by Philo Taylor Farnsworth, then twenty-one years old.

The Beaver, Utah native was just fourteen when he brought electricity to his family's cabin—which it had previously lacked!—and in a high school chemistry class reportedly sketched out the idea for a cathode ray tube that would become a key component of electronic circuits in radios and televisions. The patent he applied for incorporated not just power for the TV, but contrast and control.

How the thing works is a bit more complicated. When the TV is turned on, as electrons within the tube strike the coating of the screen and emit light, electromagnetic coils on the outside of the tube send an electron beam across the screen fast enough for the light to form continuous images.

Like many inventors, Philo Farnsworth received wide recognition only posthumously—he died in 1971, penniless, at the age of sixty-five. But he was inducted into the National Inventors Hall of Fame in 1984. Yet even that belated distinction doesn't do justice to the enormity of the change, technologically and culturally, that his invention brought to the world.

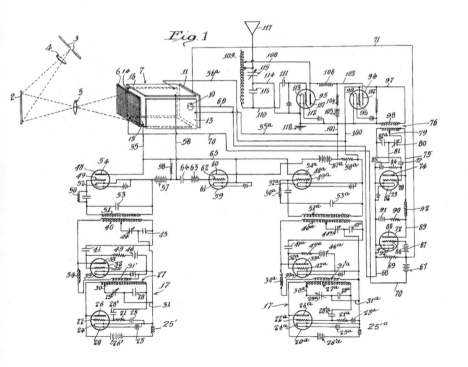

Fig. 1

## Baby Patting Machine

US3552388
Inventor: *Thomas V. Zelenka*
Patent Granted: *January 5, 1971*

We weren't parents in the early 1970s, so we can't say for certain, but judging by this patent, parents back then mustn't have had much patience with putting young ones down for the night.

The device shown here seems harmless enough: periodic pats administered by means of a bracket supporting a motor and pulley on a motor shaft that, in turn, supported an arm with a soft pad at the end. What's "objectionable," as you may see, is the inspiration for the invention.

"It is generally well-known to most parents of small infants and small children," the application's abstract reads, "that it is sometimes difficult for the infant to fall asleep, and the parent must resort to patting the baby to sleep by repeated pats upon the hind parts thereof. This can be a time-consuming operation particularly when the infant is restless . . . and it is particularly objectionable to the parent when this takes place during the night, thereby disturbing the parents' own sleep."

This is another of those patents that raises more questions than the application answers. Who exactly is, or was, Thomas V. Zelenka? How many children did he have? Where was Mrs. Zelenka and what did she think of her husband's invention, if she even knew about it? What if the baby turned over in his or her fitful sleep? How have the Zelenka children turned out? Any behavioral problems in school with, say, hitting other children on the "rump or hind parts"? If they're married and parents themselves, how do they get *their* infants to go to sleep?

On second thought, maybe it's better not to know.

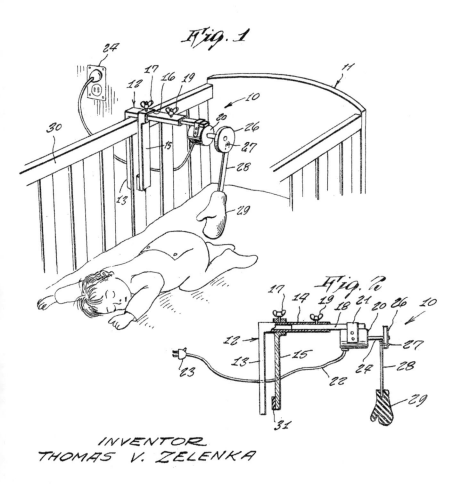

*Fig. 1*

*Fig. 2*

INVENTOR
THOMAS V. ZELENKA

## Alarm That Whacks Sleepers Awake (time-controlled mechanism)

US1293102A
Inventor: *John D. Humphrey*
Patent Granted: *February 4, 1919*

You've slapped your alarm clock silly to make it shut up. Now it's the alarm's turn. Rude awakening? Alarming? Maybe. But do you want to make it to work on time or not? Okay then, get out of bed or we'll turn this thing on!

There's a certain timelessness to the concept, based as it is on a trait in human nature: People don't *want* to get up and go to work!

Actually, Mr. Humphrey had a specific audience for his time-controlled alarm in mind: "While the device is susceptible of general employment," the abstract reads, "it is of particular utility, however, for use by deaf persons and nurses for invalids, the latter being not only awakened but frequently affected more or less seriously by the ringing of a bell, gong, or similar noise producing appurtenance."

Still, as he further notes in the abstract for his patent application, his mechanism is "of such nature as to impart a blow to an individual."

Oscar Mayer, meanwhile, has come up with a more appetizing—and maybe more effective—way to get people to throw off sleep's heavy quilt. In 2014, the company launched an iPhone app for an alarm clock. Wake Up & Smell the Bacon comes with a device that plugs into the headphone jack of an iPhone or iPad. When the alarm goes off, the device releases a vaporized bacon scent into the air. The app is free, but the subliminal message is that you have to buy real bacon (preferably Oscar Mayer bacon) to cook and eat once you're up. Otherwise, why get up at all?

As for Mr. Humphrey, we don't feel we can disturb him to ask what inspired his invention. Even his own device isn't likely to wake him from his rest: With all due respect, he's currently taking a dirt nap.

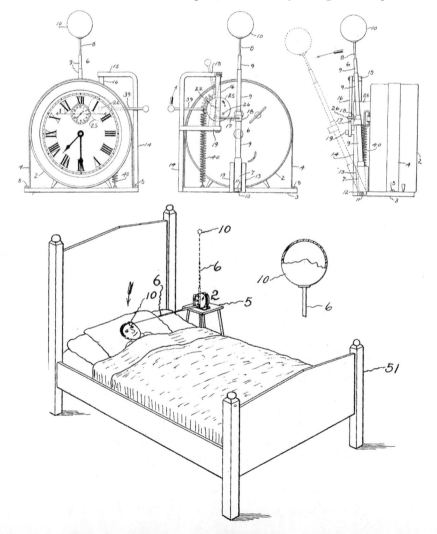

## Lawn Mower Tricycle

US4455816
Inventor: *Deanna F. Porath*
Patent Granted: *June 26, 1984*

Child labor is a global problem, with millions of young children put to work in factories and fields. But in 1982, Alabama inventor Deanna Porath sought to help solve the problem and teach children valuable lessons in innovation, multitasking, landscaping, and helping around the house.

Maybe that's stretching it a bit. In any event, Ms. Porath's invention gets kids out of the house and gets the lawn mowed. More importantly, perhaps, it shows us that a little ingenuity can go a long way to getting things done around the house!

The Lawn Mower Tricycle involves pedals that turn a front sprocket and a chain to turn a rear sprocket, mounted on a shaft equipped with a set of cutting blades. Gooseneck handlebars and a large, triangular padded seat will keep children relatively comfortable until they finish the job and are allowed back in the house for dinner (oh, sorry, that's just us spinning our narrative wheels!).

Lest readers think children's welfare wasn't taken into consideration when designing this mechanism, Ms. Porath added a protective shield mounted on the frame, "to protect the operator from debris thrown up during the course of mowing."

With its patent granted in 1984, the Lawn Mower Tricycle has since been joined by a handful of similar inventions, several of which have made it to the marketplace and onto American lawns. Get to work, kids!

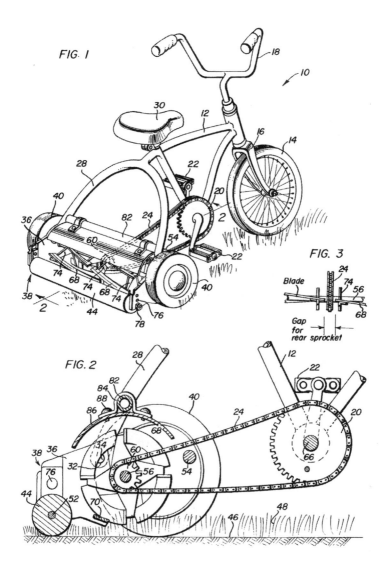

FIG. 1

FIG. 2

FIG. 3

Blade

Gap
for
rear sprocket

## High-Five Device (apparatus for simulating a high-five when watching a sports event alone)

US5356330
Inventor: *Albert Cohen*
Patent Granted: *October 18, 1994*

Some patents draw us into the minds and lives of the applicants on the day they entertain the idea for a new product or service. Had we actually been with Mr. Cohen at the time, of course, this device may not have occurred to him at all: He would have had us over to watch a sports event and high-fived us, if he insisted, after a great play. Yet that would have left us with a blank page in this book, so never mind!

The Troy, New York, inventor adroitly set the stage for his device in the Background of the Invention section of his application: "During a televised sporting event," it reads in part, "a 'high five' is commonly shared between fans to express the joy and excitement of a touchdown, home run, game-winning basket, birdie, or other positive occurrence. Unfortunately . . . a 'high five' requires the mutual hand slapping of two participants, wherein a first participant slaps an upraised hand against the elevated hand of a second participant. As such, a solitary fan is unable to perform a 'high five' to express excitement."

Improvement of hand-eye coordination, as well as jumping ability through practice are, admittedly, listed as auxiliary goals. Still, Cohen's apparatus for simulating a high five is one of those inventions and patents that raises a number of uncomfortable questions.

Aside from "why"—as in, Why did you feel the need to invent this, huh?—there are these: What if, when you go to high-five the device, it suddenly drops its hand, dissing you and making you slap thin air? Do you diss the device back by engaging it only to drop your hand at the last second?

We couldn't find Mr. Cohen to ask. Then again, we didn't try very hard; we were busy high-fiving each other because . . . we could!

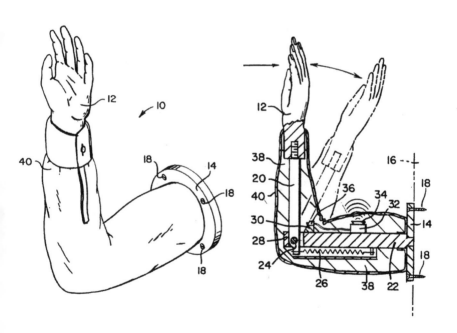

## Theater Seating Equipment (exiting through a trap door in the floor)

US151777A
Inventor: *Louis J. Duprey*
Patent Granted: *December 2, 1924*

Nineteen twenty-four was a big year for Broadway. The curtain went up on hundreds of shows making their debut on the Great White Way, including George and Ira Gershwin's first big hit, *Lady Be Good*. Theaters were packed and audiences were thrilled (or bored, depending) by shows of every variety.

But Louis J. Duprey saw all this jam-packed, sold-out success as an annoying inconvenience for the theatergoer, at least, and a potential for disaster, at worst. Everyone has experienced the frustration that comes when, seated near an aisle, you must constantly stand up to allow a fellow audience member to pass, then sit back down, then rise for the next one, ad infinitum. But that wasn't all that was vexing Duprey. He was concerned about the panic that often ensued during a medical emergency or fire. How could the entire audience be swiftly and simultaneously evacuated?

Duprey envisioned a mechanical solution. Every seat in the theater would be mounted onto an individual apparatus that would, on command, lower the seat into a custom-built, fireproof subchamber, from which the bewildered audience members could "conveniently" exit. What about the mezzanine and balcony seats, you may ask? Don't ask.

Although a noble idea, this system was impractical, to say the least. Imagine selling the concept to a theater owner. In order to retrofit an existing space, the entire orchestra seating floor would have to be removed to access the "subchamber," if there was one. Then hundreds of seats would have to be connected to hundreds of electric motors and hundreds of mechanical lowering devices.

Oh, and one other minor detail: The cost of this wildly complicated conversion would raise the price of a ticket to see a show to that of a trip to London.

A time when going to a show could be a life-threatening experience.

## Toilet Tank Aquarium

US5983411A
Inventor: *Herbert Demoret*
Patent Granted: *November 16, 1999*

A thirteenth-century English Franciscan friar named William of Ockham, in an effort to defend the idea of divine miracles, came to the conclusion that "Entities should not be multiplied without necessity." In other words, the simplest solution is usually the best one. Over the centures, this problem-solving principle became known as Ockham's razor.

In 1998, Herbert Demoret somehow discerned a need for a toilet that would serve as both an aquarium and an expeditious way to swiftly deliver any deceased piscine occupants back to their final reward in mother ocean. However, in Demoret's patent application, this simple idea, in a stunning reversal of Ockham's thesis, managed to complicate the idea almost beyond recognition. In the process, a record may have been set by using the word *assembly* eleven times in one paragraph.

"A new toilet tank assembly aquarium for housing aquatic creatures. The assembly includes a toilet bowl assembly with a toilet tank assembly coupled to the toilet bowl assembly. The toilet tank assembly has a top, a front, a back, and a pair of sides. The front, back, and sides of the toilet tank assembly define an interior space with the top of the toilet tank assembly having an opening into the interior space of the toilet tank assembly. A toilet water reservoir is located in the interior space. A lid substantially covers the opening of the top of the toilet tank assembly. The front of the toilet tank assembly is generally transparent."

Sorry, Friar Ockham.

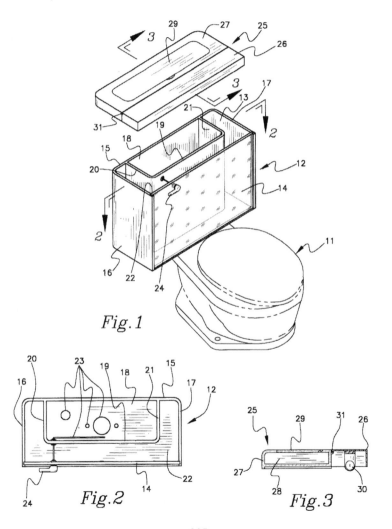

Fig. 1

Fig. 2

Fig. 3

## Birthday Cake Candle Extinguisher

US3168983A
Inventor: *Paul Bosak*
Patent Granted: *February 19, 1965*

Over the course of our lives, certain sights, sounds, and smells mark the passage of time. From the sweet scent of holly boughs at Christmas, to the spooky glow of a jack-o'-lantern on a crisp Halloween night, or the thrill of that first bike ride with no training wheels—these are the memories that fill the pages of our personal scrapbooks.

But, for any kid, one event towers above the rest—his birthday. On this one day of the year, all the attention is lavished upon him. Hugs and kisses abound, presents arrive from every direction, and, to top it all off, the most beautiful cake in all the world, with its constellation of glittering candles, is set before the birthday boy or girl, usually accompanied by a shockingly out-of-tune rendition of "Happy Birthday."

Finally, the big moment arrives. First, make a wish, don't tell anybody, then take a big breath and, with earnest determination, blow out those candles—or at least try to.

Inventor Paul Bosak probably saw this joyful celebration a different way. This was a ritual fraught with opportunity for catastrophe. So, with fire injury and unsanitary expectoration looming large in his mind, he conceived a mechanical solution to address these potential disasters. His device comprised a vertical, hollow, tubular sectional body on a base. The top section was formed in the shape of a human face with an aperture in the mouth through which compressed air from an internal can would be discharged, thereby blowing out the candles remotely and avoiding the situation in which children inadvertently spit on the cake.

"Happy Birthday, Emily! Now go put on your goggles and hazmat suit. It's cake time!"

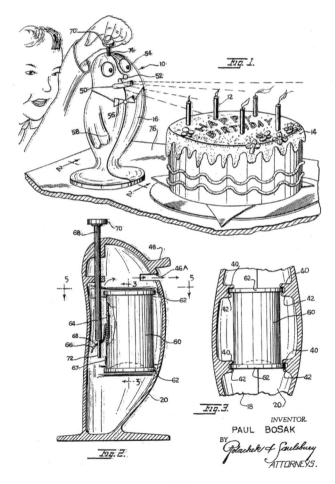

*Fig. 1.*

*Fig. 2.*

*Fig. 3.*

INVENTOR.
PAUL BOSAK
BY
*Polachek & Saulsbury*
ATTORNEYS.

## Microwave Oven

US2495429A
Inventor: *Percy Spencer*
Patent Granted: *January 24, 1950*

Percy Spencer was a kid from a little town in Maine who, like so many in the post-Depression years, was forced to go to work at an early age to support his family. The paper mill where he worked was about to begin using electricity to power itself, a groundbreaking new concept in his rural home region. Fascinated, Spencer set out to learn as much as he could about this "invisible" source of power. Thus, he was one of three people hired to install electricity in the plant, even though he never finished grammar school.

Later, after joining the Navy at the age of eighteen, he taught himself radio technology, trigonometry, calculus, physics, and metallurgy—while he was standing watch at night! The guy was a full-blown nerd, a geek, and a brainiac before such monikers existed.

In 1945, while employed by the defense contractor Raytheon, he noticed that the microwaves from a radar set he was working on had melted a chocolate bar in his pocket. Unwittingly, he had come upon a solution without a problem. But now, with a raison d'être, it wouldn't be long before the microwave oven came to be. And, in a nod to the god of irony, the first food ever deliberately cooked in a microwave was popcorn. The second was an egg, which promptly exploded.

In 1945, Raytheon filed a US patent application for Spencer's microwave cooking process, and the first unit was soon placed in a Boston restaurant for testing. It was six feet high, weighed about 750 pounds, and cost $5,000. In 1967, the first relatively affordable ($495) countertop microwave oven hit the market. Early adopters soon realized that "nuking" that cup of tepid coffee for just seconds too long would turn the liquid into something akin to volcanic lava. And drinking it would result in a scalded tongue, prompting a string of epithets sure to shock anyone within earshot.

But buttery popcorn in two minutes? Well, that was worth the wait.

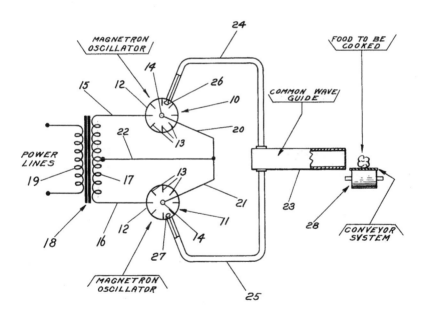

## Dishwashing Machine

US355139A
Inventor: *Josephine Cochrane*
Patent Granted: *December 28, 1886*

These days, most people, married or unmarried, find the term *housewife* insulting, or at least tone-deaf, because it defines an occupation in terms of a woman's relation to a man. and her role as a housekeeper. In Shelbyville, Indiana, in 1886, Josephine Cochrane, a newly minted socialite, married to a successful merchant, saw the world through a more independent lens.

The Cochrans (Josephine added the "e" to her married name because she felt like it, okay?) had a busy social life. Today, they'd be the "power couple" on the internet. In 1870, the couple moved into the perfect house for entertaining with servants, real glass windows, indoor plumbing, and all the modern conveniences. There, they threw dinner parties that were the envy of all the other folks who didn't know what a convenience was. Josephine loved pulling out the china from the 1600s—which actually came from China. After one such event, the "help" carelessly chipped some dishes. This raised the ire of Cochrane, who henceforth took on the dishwashing chores herself, suffering what may have been the first identifiable case of "dishpan hands."

That's when she had the idea. This was the late nineteenth century—if someone could invent a machine to sew clothes and cut grass, then how hard could it be to mechanically wash the dishes? And so, the lady of the house went out to the back shed, and soon proved her mettle and ingenuity with a prototype machine that, not surprisingly, worked.

After receiving a patent in 1886, her factory business began production in 1897. The manufacturer that produced the early machines would later be known as Kitchen-Aid.

Today, the fight isn't over who's doing the dishes; it's about who stacks the dishwasher correctly.

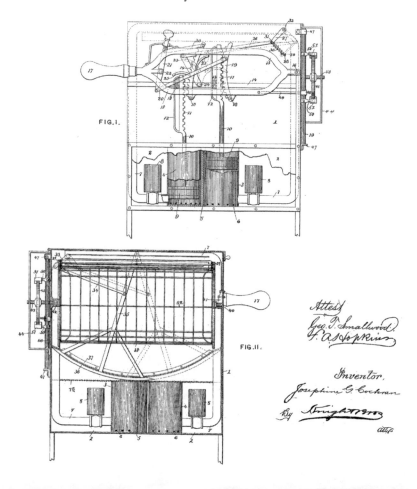

## Jones Air-Conditioning Unit

USD132182S
Inventor: *Frederick McKinley Jones*
Patent Granted: *July 12, 1940*

It's a gray winter's day. Outside, the snow is falling and a chill wind blows. But when it's time for dinner, we have the luxury of choosing from a veritable cornucopia of fresh produce, poultry, or meats stored conveniently in our refrigerators. We don't give a second thought as to how it is we can enjoy a juicy apple from Washington—in Miami on a hot afternoon in July—or a tender New York strip steak in Nome, Alaska during a blizzard.

For these impressive quality-of-life improvements, we have American inventor Frederick McKinley Jones, the inventor of the first automatic refrigeration system for long-haul trucks and railroad cars, to thank. Jones was born in Cincinnati, Ohio on May 17, 1893, to a white father and a black mother. Deserted and virtually orphaned by his parents as a child, he was raised by a priest at a Catholic rectory. Showing an early aptitude for tinkering with machines, at age fourteen he began work as an automobile mechanic. Then, his natural curiosity blossomed into a lifelong desire to seek new challenges, fueling his search for advancement against the steep racial odds of the time.

In 1935, he invented a roof-mounted cooling device for trucks and railroad cars. This groundbreaking new system all but eliminated the risk of food spoilage during long-distance hauls and was later adapted to a variety of other common carriers, including ships and railway cars. These successes led to the formation of the Thermo-King Corporation, which revolutionized the field of transport refrigeration.

Portable cooling units designed by Jones were critical during World War II, preserving blood, medicine, and food for use at army hospitals and on open battlefields. In 2009 Frederick was dubbed the King of Kool by the trucking industry and his technological breakthrough redefined the global marketplace, with cultural reverberations felt from the world's largest cities to its most isolated villages.

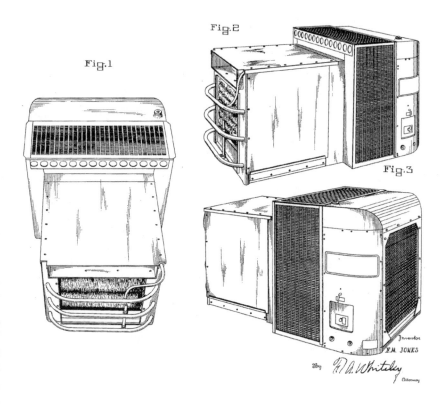

Fig.1

Fig.2

Fig.3

Inventor
F.M. JONES

By R.A.Whitely
Attorney

## Telecaster Guitar

USD164227S
Inventor: *Clarence Leo Fender*
Patent Granted: *August 14, 1951*

What do Bob Dylan, Vince Gill, David Gilmour, Merle Haggard, George Harrison, Chrissie Hynde, and Keith Richards have in common? You'd be right if you said they're all famous musicians. But there's something else. Passing through the glittering careers of every one of these stars, a unique musical legacy has been carried forward. They all played a Telecaster guitar. And there's one man to thank for creating the instrument that has kept the world rocking since it burst upon the scene in 1951: Leo Fender, a man who couldn't play a lick on guitar.

He did, however, have a natural talent for electronics, and, as it turns out, was graced with impeccable timing. After World War II, when the Big Band sound was on the wane, small combos playing boogie-woogie, rhythm and blues, and western swing brought "the beat" to a nation hungry for musical excitement. These hot little bands had embraced the newfangled electric guitar, but there was a big problem. The early electric guitars were big, hollow archtops, which, when amplified even a smidge too loudly, would "feed back" in an uncontrollable, ear-splitting howl.

But with the increasing popularity of local dance halls and roadhouses, a need arose for louder, more durable instruments. Leo Fender intuitively saw a potential market for an electric guitar that would stay in tune, was easy to play, and, most importantly, wouldn't feed back when played at dance hall volume. In 1948, he completed a prototype for a thin, solid-body electric. It was first released in 1950 as the Fender Esquire with one pickup, then renamed the Broadcaster and finally the Telecaster, featuring two pickups. That basic design is still being produced today and remains at the top of the list of the world's most popular musical instruments.

But, back in the rockin' fifties a new phrase, usually uttered by an exasperated parent, entered the lexicon for the first time: "Turn that damn thing down!"

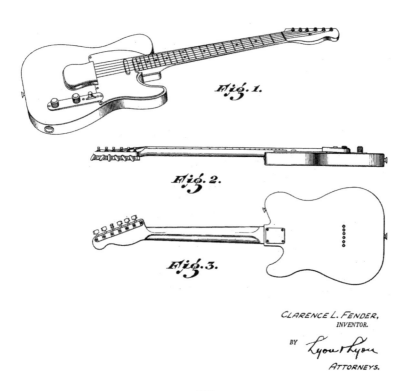

*Fig. 1.*

*Fig. 2.*

*Fig. 3.*

CLARENCE L. FENDER,
INVENTOR.

BY

ATTORNEYS.

# ACKNOWLEDGMENTS

Let us first acknowledge the legions of inventors and dreamers, some of whose flames burned more brightly than others, but all adventurous souls, without whom this book wouldn't be.

Thanks to Jordana Hawkins, our editor at Running Press, whose generous and gracious support enabled us to do what we've done; our talented book designer Rachel Peckman; and our copy editor Diana Drew for her meticulous work. Early advice and support came from long-time patent attorney and friend Paul Fattibene and from Alan Neigher, who has championed our projects for many years and in many ways. Sean Patrick Suiter, founder of the intellectual property law firm Suiter Swantz, opened his capacious and well-furnished mind to us.

Thanks, too, to the United States Patent and Trademark Office in Washington, DC, where patent applications are not only processed, but their copy and illustrations made public for all to read, publish, and be inspired by.

Special thanks are due David Connor, Norman Lear, and Robert Downey Jr. for their unflagging support and Lisa Grenadier and Jackson Connor for all that they are.

Additional appreciation goes to our research associate Ryan DeWitt, who, though still in high school, came to our rescue at critical times, and to his father, Chris DeWitt, for his support and technical advice.

Thanks, finally, to Rita Kurs, who has a bookful of patentable ideas in her head and to Maggie Smith, Dean Builter, and Elizabeth Hilts.

# ABOUT THE AUTHORS

**Tom Connor and Jim Downey** are the coauthors and producers of a dozen and a half books, as well as the authors of the national best-selling parodies *Is Martha Stuart Living?*, *Martha Stuart's Better Than You at Entertaining*, and *The Smyth 'n Hawk'em Gardening Catalog*.